About Island Press

Island Press is the only nonprofit organization in the United States whose principal purpose is the publication of books on environmental issues and natural resource management. We provide solutions-oriented information to professionals, public officials, business and community leaders, and concerned citizens who are shaping responses to environmental problems.

In 2000, Island Press celebrates its sixteenth anniversary as the leading provider of timely and practical books that take a multidisciplinary approach to critical environmental concerns. Our growing list of titles reflects our commitment to bringing the best of an expanding body of literature to the environmental community throughout North America and the world.

Support for Island Press is provided by The Jenifer Altman Foundation, The Bullitt Foundation, The Mary Flagler Cary Charitable Trust, The Nathan Cummings Foundation, The Geraldine R. Dodge Foundation, The Charles Engelhard Foundation, The Ford Foundation, The German Marshall Fund of the United States, The George Gund Foundation, The Vira I. Heinz Endowment, The William and Flora Hewlett Foundation, The W. Alton Jones Foundation, The John D. and Catherine T. MacArthur Foundation, The Andrew W. Mellon Foundation, The Charles Stewart Mott Foundation, The Curtis and Edith Munson Foundation, The National Fish and Wildlife Foundation, The New-Land Foundation, The Oak Foundation, The Overbrook Foundation, The David and Lucile Packard Foundation, The Pew Charitable Trusts, The Rockefeller Brothers Fund, Rockefeller Financial Services, The Winslow Foundation, and individual donors.

Reaping the Wind

To William Heronemus and all those brave enough to speak out about the folly of our current energy system.

Reaping the Wind

How Mechanical Wizards, Visionaries, and Profiteers Helped Shape Our Energy Future

Peter Asmus

ISLAND PRESS
Washington, D.C. • Covelo, California

Library of Congress Cataloging-in-Publication Data

Asmus, Peter.
Reaping the wind : how mechanical wizards, visionaries, and profiteers helped shape our energy future / by Peter Asmus.
p. cm.
Includes bibliographical references and index.
ISBN 1-55963-707-2 (cloth : alk.paper)
1. Wind power plants. 2. Electric utilities—United States. I. Title.
TK1541 .A75 2001
333.9'2'0973—dc21

00-010901

Printed on recycled, acid-free paper 

Printed in Canada
10 9 8 7 6 5 4 3 2 1

Contents

Acknowledgments

I began this book in 1991 under the naïve notion that it would be completed within a year or so. I never dreamed it would take so long to complete, but my life, which took several twists and turns, intervened. The basic themes of the book kept evolving as I researched and wrote, and as the wind power industry's fortunes rose and fell—and then rose again. Looking back, however, the timing for a book on wind power at the onset of the new millennium could not have been better.

First off, I would like to express my sincere thanks to Todd Baldwin, senior editor at Island Press, who persevered through numerous drafts and helped pound this complex story into a single narrative that does justice to the story of American wind farming.

There are several wind farming industry characters I met in the early 1990s that were critical in convincing me there was plenty of rich human drama in the wind farming industry to produce a riveting book. Two people in particular stand out: Randy Tinkerman and Jim Dehlsen. Without their perspectives and persistent encouragement, this book might never have been written.

Over the course of the last decade, many other individuals were invaluable compatriots, opening up their immense files, putting me in touch with some of the old-timers, or just offering great anecdotes or other tidbits. These include Robert Kahn, who loaned me his entire wind power clip files when I first began my research. His efforts were complemented by the following individuals: Woody Stoddard, Don Smith, Dick Curry, Jim Williams, Ty Cashman, Randy Swisher, Phil Leen, Paul Gipe, and Robert Thayer. Jack Hicks, and a bevy of writers that included Gary Nabhan, all helped critique early chapter drafts during the "Art of The Wild" writers' workshops and infused the book with a sense of place.

Since this book was written over such a long period of time, bits and parts were supported —often indirectly—by a variety of organizations. These include the William and Flora Hewlett Foundation, the Center for Energy Efficiency and Renewable Technologies, the California Energy Commission, the Natural Resources Defense Council, and *Terra Nova.*

Finally, I would also like to offer immense thanks and special gratitude to Leslie Lipman, whose unwavering support during crunch times freed me to relax when I really needed to take a break from the intense schedule required during the final two-year push to complete the manuscript. I also extend heartfelt appreciation to Lori Ann, who was there when it all began and offered critical and continued encouragement during my first encounter with writer's block. A special thanks to Debbie Jolly, John Sibbet, and Jeremiah Adams for helping me free my creative spirit. There are several others who readily offered boundless enthusiasm. You know who you are. Last, and certainly not least, I offer thanks to my family members, especially Mom and Dad, who never stopped me from pursuing the insane profession of being a writer.

Prologue

The Secret World of Electricity

> The moment man cast off his age-long belief in magic, Science bestowed upon him the blessings of Electric Current.
>
> —Jean Giraudoux

Most Americans give electricity hardly a passing thought, let alone wonder what sources produce the power that in the past century forever changed the way we live. Electricity's engineering and logistical details remain a mystery to most of us. This elixir of the metropolis is indeed a one-of-a-kind product. It is everywhere yet is nowhere. It is virtually unseen, and the stuff has always been supplied by a faceless local utility.

Electrons, those infinitesimally small subatomic particles holding a negative charge, spin in orbits around the larger, positively charged protons at rest in the nucleus. When electrons are torn from atoms, by friction, induction, or chemical changes, we have electricity. Electric current is so named because it is a flow of electrons in a circuit. However, the electrons themselves do not literally move over great distances; electric current acts more or less like a force field, a sort of vibration of energized bits that can be channeled over wires and into the plethora of appliances, computers, and telecommunications tools that now dominate the interiorscapes of our homes in the twenty-first century.

The "discovery" of electricity might be said to date back to around 600 B.C., when the Greek philosopher Thales made the first known recorded observations of static electricity. Electricity has a rich history of theory and experiment, particularly in the past 300 or so years. Otto von Guericke built a generator in 1660, at the same time demonstrating that electricity could be transmitted. In the eighteenth century, the people theorizing, investigating phenomena, and building experimental devices were too numerous to name. Interestingly, the first "practical" applications of electricity were developed by doctors who prescribed it as a tonic for treating whooping cough and diseases of the skin. This strange force is now ubiquitous in the industrialized world.

Distributed on a "grid" of high-voltage lines, transformers, and relay stations that connect most homes and businesses and other institutions to sources of power that can be located several hundred miles away, electricity has had an ever greater influence over our lives—especially since the advent of the digital age. Still, the only time we think about electricity is when the power actually goes out.

Electricity is a phenomenon that is still hard for most people to get their minds around. Not only has it been the preoccupation of an orphic society of geeky engineers and crafty entrepreneurs such as Thomas Edison, but the business details of how this mystery product has been bought and sold have largely been kept under wraps since the turn of the past century. Utilities assumed that demand for this stuff would march in lockstep with an ever-expanding population and booming economy. This fundamental assumption held sway until the 1970s. Then steep recession dampened expected economic growth and in turn vastly reduced demand for electricity. Suddenly, there was excess capacity, and the new supply coming on-line was costing far more than anticipated, primarily because of construction cost overruns at nuclear power plants, such as California's Diablo Canyon. Environmentalists then began to argue successfully for reducing demand, instead of building more power plants, by boosting efficiency through energy-saving lightbulbs, appliances, and housing stock.

A second response to the ever-increasing costliness of utility-supplied power was to allow private companies unaffiliated with the monopoly to build smaller power plants driven by fuels other than the coal and radioactive isotopes on which the utilities mainly relied. Jimmy Carter initiated a trend toward privatization and competition, and California most aggressively implemented this new vision. It was Governor Jerry Brown's administration that set the stage for integrating smaller, more diverse, power generators into the electric grid. Sure, there were rural regions not connected to the grid. But the vast, vast majority of consumers were plugged into a power supply system that was, for the first time, being opened up to new players.

Wind power was in the lead of this new movement toward competition, toward shedding some light on our energy future, toward shifting from larger to smaller, from the familiar to the new and cutting edge. Still, there was usually nothing the average person could do about where his or her power came from—or what it cost. The utilities reluctantly bought power from wind farmers, but the consumer's sole choice was to purchase electricity from a mix of fuels still dominated by coal, nuclear, and large hydroelectric power plants.

This was the situation in most communities, though not in Sacramento, California. I began to learn about electricity during an energy battle right in my own backyard. The Rancho Seco nuclear power plant had made the national news in 1987 because a long list of problems had provoked local rate increases exceeding 200 percent. The battle to close down this nuclear lemon had begun, and I was hired to cover it for the national energy trade press. There were rumors of drug use, and even sex orgies, under the immense cooling towers. Some insiders painted a picture of an operations crew that behaved like a bunch of yahoo cowboys straight out of an episode of *The Simpsons* TV show.

My new job was to report on the Sacramento Municipal Utility District—my local publicly owned electric utility with the unappealing acronym SMUD, the owner of Rancho Seco. Here was a situation where local citizens wanted to close a nuclear reactor and had collected enough signatures within the community to force not one but two public referendums. One of them was passed, and ratepayers were then asked to vote on what sources of power should replace Rancho Seco. A community consensus revolved around the use of cleaner sources, including renewables such as wind power.

Over the course of the next 10 years, I learned the ins and outs of the electricity business, the world's largest—and most polluting—industrial enterprise. To outsiders the industry is boring and complex, which explains the widespread historical ignorance about its extremely important activities: More capital has been poured into the building of power plants and associated infrastructure than into any other industrial activity. For decades sleepy state regulatory commissions gave the monopolies virtual carte blanche in planning and building plants, approving multi-million and multi-billion dollar decisions with little scrutiny of their economic or environmental impacts. The consequences of those decisions, and the fact that government subsidies helped promote the fiction that they were cost effective, are the reason that world leaders must come up with an accelerated program for renewable energy deployment.

It wasn't until the Energy Policy Act of 1992 was passed that the grid itself was opened up to allow competing suppliers to sell back and forth across transmission lines previously made unavailable by steep transport fees. This new law set the stage for a new electric generation and distribution regime in which all sellers would pay the identical rate to transport their power via the grid. The host utility could no longer use its control of the grid as a way to block competing utilities and other suppliers from making power sales.

The consequences of this change are now being seen across the country, as one state after another passes laws dismantling its public utility monopolies. One outcome of this wave of deregulation is that growing numbers of consumers will soon be able to choose their particular power suppliers just as we now choose telecommunications companies. This has opened the door to electricity from popular alternative energy sources, long suppressed. Today, wind turbines are cropping up in many parts of the country, dumping their electricity into grids and earmarking it for those consumers who pony up to purchase green power.

Wind power has suddenly become a credible industry. Worldwide capacity has increased by about 25 percent per year since 1990, jumping from 1,700 MW a decade ago (most of it in California) to 15,000 MW today—equal to the output of about a dozen nuclear reactors. That capacity is expected to more than double over the next five years; Germany alone has tripled California's total wind power production in less than five years. And the utility monopolies that once sought to kill the industry are now among its prime investors. During the past 20 years, wind farming emerged from a dream of a few to an industry that is now the fastest growing source of electricity in the world.

Still, the world's wind resource is largely untapped. For example, the winds blowing on just 6 percent of the windiest land sites in the United States (excluding Hawaii and Alaska) could supply one and one-half times the entire nation's electricity needs. There is plenty of wind available to reduce our reliance on fossil fuels. Wind power will need to play a major role if we are to stave off catastrophic global climate change. If the current balance among modes of power production continues, and at current levels of per capita consumption, our addiction to electricity could collapse a global climate system that has sustained life for millennia. And the onset of this global catastrophe will only be accelerated if that balance is spread throughout the developing world, whose population exceeds 2 billion and is continuing to grow.

The key obstacle to widespread deployment of wind power is cost, but this barrier to entry is slowly eroding. Many experts speculate that wind power will remain the fastest growing power source well into the new millennium. Costs have dropped 80 percent since the first installations in California in the '80s. Prices are projected to drop another 20 to 40 percent over the next 10 years.

Perhaps the most surprising endorsement of this technology, frequently associated with fringe characters and fly-by-night firms, came in 1999: The U.S. Department of Energy inaugurated "Wind Powering America," a strategic plan designed to provide 5 percent of our electricity from wind power by the year 2020—which equates to 80,000 MW of new capacity. Never before has the federal government set such an aggressive target for private-sector initiatives on renewable energy. Given all the questions about whether the technology would ever work, all of the outrageous tax scams during the start-up years in California, and then the embarrassing collapse of the world's largest U.S.-based wind farmer in the mid-'90s, a strong case can be made that wind power's current popularity is the outcome of the most improbable success story in the history of technologies that produce power.

That story is chronicled in this book. It is a story with deep historical roots, which can be traced back to medieval times and earlier. But the center stage is the twentieth century and the Altamont Pass, just east of San Francisco. The tale is in part a drama, one of high-profile success and downfall, most notably that of the industry giant U.S. Windpower (later renamed Kenetech), which first arrived in the Altamont in 1980. But the story extends well beyond the Altamont, merely the most visible of the many wind power sites that were developed in the 1980s in California. It is also the story of Zond (now Enron Wind Corp.), Kenetech's chief rival in the 1990s. And it is about Fayette, a company hatched by a CIA boss and U.S. Windpower's first adversary in the Altamont Pass. Most of all, however, this book is the story of people like Mike Bergey and Randy Tinkerman, Woody Stoddard and Terry Mehrkam, and dozens of other visionaries, iconoclasts and dreamers, schemers and true believers, and mechanical wizards and profiteers—all of whom helped transform the way energy is produced and distributed in the United States today. For as spectacular as its failures have been, the wind industry is seemingly finally on the brink of a new era—defined as much by its large and well-known purveyors, like Kenetech and now Enron Wind Corp., as by the hundreds of much smaller fry—the Tinkermans, Bergeys, Thomases, Carters, Sherwins, Cashmans, Leens, Harmons, and Lees—who are still waiting.

Part I

Chapter 1

Birds of a Feather

Why is betrayal the only truth that sticks?
—Arthur Miller

"I hear the day is this coming Monday," blurted somebody into my phone answering machine, the surly voice edged with a New York jeer. "It's dumpster diving time," he sang as if reciting lyrics to some popular rap tune. It was Randy Tinkerman, I quickly surmised. A short man with a long, salt-and-pepper ponytail, Tinkerman has really big ideas about energy that were shaped by the 1970s, a time when an individual really seemed to be able to make a big difference in the world. There was revolution in the air, and Tinkerman was one among the warriors seeking to alter dramatically the way we power society.

His phone message referred to his unrelenting search for documents that might prove valuable in courts asked to pass judgment on some pretty wild promises of profits to be made from the wind. The target of inquiry was Kenetech, formerly known as U.S. Windpower. Tinkerman calls it "Wind Potato," a moniker reflecting his profound lack of respect for what had been America's largest wind farmer. Monday was the day Kenetech was going to make it official. It was going out of business, yet another company to go under while trying to strike it rich, and purportedly to save this precious planet.

Though Tinkerman is a long-time advocate for wind power, he has never been afraid to speak up, or sniff around garbage cans, when he has detected outrageous claims for the technology—or felt that one of his close friends within the industry has been wronged. The harsh demands of large-scale corporate wind farming can bring folks together. They can also tear former allies apart.

Tinkerman has been involved in more than a few legal disputes with Wind

Potato. One such dispute was a disagreement about the actual amounts of energy produced by wind farms in the Altamont Pass. The corresponding payments to the landowners, mostly old ranchers, for the use of the wind that blew on their land were based on deficient data, Tinkerman claimed. He lost a critical case against Kenetech that would have awarded the ranchers more royalty revenues from their wind rights—a surprise verdict since the case seemed to be going his way until the very last minute. The verdict reeked of foul play, claimed Tinkerman, who speaks with an air of great authority even when indulging in the most sinister of conspiracy theories.

From the message on my answering machine, it seemed clear that Tinkerman felt vindicated by the imminent collapse of his object of wrath. All of the warnings he issued about the company for all of those years now appeared to be right on the mark. Sure enough, Kenetech filed for Chapter 11 bankruptcy protection on May 29, 1996. It was unlikely the company would ever do business again. For his part, Tinkerman feared he would lose his trail of evidence once the company folded, which accounted for the urgency in his message. His bird's-eye view of the wind industry landed him consulting gigs with highly reputable law firms as well as wind companies and utilities, and he ended up drafting large portions of a shareholder lawsuit that would charge that Kenetech was guilty of fraud (though the suit was dismissed by a judge in the U.S. District Court in August 1999 but has been appealed to the U.S. Court of Appeals, 9th District).

The fall of Kenetech was a watershed event. Just at a time when the wind industry seemed on the verge of respectability—with Kenetech promising new wind turbines that could compete, without public subsidy, with traditional electricity-generating technologies—the flagship firm of the domestic wind power industry suddenly was no more. Not only had the company declared bankruptcy, but it apparently had engaged in questionable business practices, according to Tinkerman and many of its competitors.

Tinkerman's glee, and wistful melancholy, over the bankruptcy springs from a complex web of stories revolving around a company that once inspired him like no other. The dirty laundry of the industry seemed both to repel and compel him. He still spoke fondly of the ideals of the original visionaries, the gusto exhibited by dozens of wind energy pioneers, some of them present at the founding of U.S. Windpower. Nonetheless, the dry but critical details of the electricity business, and the odd currency of the wild and woolly wind economy, conspired to transform this story of hope into one of betrayal. That was the gist of what Tinkerman told me, if you boiled down his streams of meandering and often caustic comments. Betrayal not only of the industry's original ideals, but of some of his closest friends who first championed the wind power cause. Wind Potato, in Tinkerman's eyes, was guilty of letting him, and a whole cadre of small, struggling wind farmers, down.

◆

Tinkerman, whose education never extended beyond high school, came to the wind industry because he was desperate for a day job. Born in Chicago but raised in New York's Hudson River Valley, Tinkerman always believed that the way to get things done is to jump in with both feet. While he used this approach to become a country and swing session drummer, he still needed steady money for the basics: food and housing. He was driving home one day in the early 1970s after a disappointing trip to an employment agency in Kingston, New York. He had donned his best shirt and delivered a hand-written resume. "You're kidding," the agency representative said, shaking his head in disbelief. Dejected, Tinkerman decided he would need to take matters into his own hands. Coming upon a shop that sold snowmobiles, chain saws, lawn and garden tractors as well as stoves and fireplaces, Tinkerman pulled into the parking lot. He told the owner, Paul Sturges, he had been a farmer and "there is nothing I can't do." Sturges, who just happened to be a solar energy pioneer in the 1950s, was a wild man aristocrat, a cross between Jimmy Stewart and Albert Einstein. He liked Tinkerman's attitude and told him he would give him a call if he needed him. A week later, Sturges called and offered Tinkerman a job. "He wanted me to be the chief engineer on the most sophisticated heat recovery system for burning wood ever invented. Sturges had been working on it for years. I couldn't even spell the word engineering!" said Tinkerman.

With his hair flowing well down his back and a full beard hiding virtually all of his face, Tinkerman became just another hippie wanting to learn more about wind and other renewable energy technologies. "I wanted to disconnect from the grid and didn't want to use poison as my source of power," said Tinkerman. He fondly reminisced about smoking dope with icons of the nation's aristocracy and fantasizing about a clean-energy future. "I had lived in a teepee for a number of years and so was quite comfortable with the idea of living close to nature. I was totally into the 'tune in, turn on, and drop out' creed espoused by Timothy Leary, and I thought wind power was a really cool alternative to drop into. I began to recognize that wind power was a technology that allowed a healthy lifestyle that would allow me to explore consciousness, in all of its forms, and the immense pleasures of being human." Tinkerman soon became convinced that wind power was *the* technology he should focus on. "It was the first solar energy technology that capitalism could feed on," he quipped. "Do you really think that Merrill Lynch believed wind power was an important technology of the future? Of course not. They were in on it because there was money to made—and lots of it."

Tinkerman attended the "Toward Tomorrow Fair," which was being held at the University of Massachusetts in Amherst, where William Heronemus, a gruff former designer of the nation's nuclear submarine fleet, was teaching a new class about the engineering fundamentals of wind power. The fair was being billed as the first "sustainable Woodstock." There Tinkerman met Woody Stoddard, an engineering student of Heronemus's. Tinkerman and Stoddard became friends. "It became clear to me that electricity was going to be my game. Since I wanted to say 'no' to the nukes, I needed something to say 'yes' to. It was Woody who began to fill me in on the nuts and bolts of wind energy systems," Tinkerman recalled. Though not enrolled in the graduate program because of his lack of qualifications, Tinkerman nevertheless began to hang around the campus. He began accumulating in-depth knowledge about designing wind turbines, the various stresses placed on the machines, and the theories and techniques envisioned to respond to these stresses. He always believed in the concept of learning through hands-on experience. Tinkerman helped Stoddard with the molds for the nacelle on a new kind of wind turbine he was designing for Heronemus's wind engineering class. Soon thereafter, Tinkerman was employed by a company called Wood, Wind and Water. He traveled as far as Wisconsin to install a wind turbine and also helped build a solar greenhouse the size of a football field in western Massachusetts.

He shows me two pictures. In the 1976 shot, he looks like a hippie and the wind turbine he's examining looks tiny and fragile. One year later, Tinkerman has shaved off all of his hair and the wind turbine he is working on is much more robust, sturdier. "These two photos sum up where the wind industry was going. I began to realize that the vehicle for changing the world was going to be business and that I would need to make a good impression upon the financial community. At the same time, wind power technology was getting bigger and better, too."

Tinkerman began meeting other wind power technologists, hitchhiking from one meeting to another, finally flying out to California in 1977 for a solar energy conference. Tinkerman decided to stay in the Golden State when offered a job that unfortunately never panned out. He needed a credible platform for his energy consulting gigs so he created "The Transition Energy Project Institute" the next year and gave himself the title of executive director for an organization that he would tell friends was his "bedroom think tank." He bought his first nice three-piece suit and began to preach to anyone who would listen about the vast possibilities available for generating electricity from the wind. "We could put wind turbines off the coast of San Francisco," he told the *California Journal*, a well-respected publication read by policy wonks and politicos, echoing an idea that Heronemus first floated after becoming disillusioned with the concept of underwater nukes. "Wind power, as well as other renewable resources, could easily supply large portions of California's electricity needs," proclaimed Tinkerman, enjoying his new platform.

But Tinkerman's sudden thrust into the limelight began to affect his sleep. He had met an artist at a cafe in North Beach and asked him to come up with an illustration of these fantastic wind power ships. He then gave the drawing to a UPI reporter. "It was a very stylized picture. It didn't matter that the wind turbine in the drawing violated the laws of physics," chuckled Tinkerman. "The drawing was so visually stunning that it captured the imagination of the reporter." The story was going to be trumpeted around the world, but the day before it was to be published Tinkerman tossed and turned, worrying whether anyone would challenge his credibility because of this spurious turbine design.

The end result? Tinkerman received a phone call from Governor Jerry Brown and in 1979 was named to a new task force on wind energy. Six months later, Tinkerman was tapped to do some real paying work on behalf of a demonstration wind power project for the California Energy Commission, a state agency that would play a major role in promoting wind power. He proposed major modifications to the original project. Tinkerman substituted a wind turbine designed and manufactured by a Texas father and son team, Jay Carter Sr. and Jr.; it was believed to be the state of the art of American lightweight machines. He also shifted the site's location from an isolated swamp to the top of a highly conspicuous hill right alongside Interstate 80, a heavily traveled highway that carried politicians and policymakers back and forth between San Francisco and Sacramento, the state capital. Tinkerman saw the turbine as a public relations ploy designed to showcase this new way to generate electricity. He crossed his fingers and hoped the damned thing would actually work.

In 1980, because of the efforts of people like Tinkerman, the California state legislature passed a set of tax credits for wind power the likes of which had not been seen before, or since. Wind farming in the state became an irresistible business proposition. The Great California Wind Rush was on, and Tinkerman joined the fray. He was one among many who tried to secure the legal rights to the winds that blew across the barren Altamont landscape. But it wasn't until 1984 that Tinkerman finally got his piece of the action there. He brought in a British company to develop wind turbines in the 300 kW range—six times the size of the original 50 kW USW machine—and ultimately a wind turbine that was 750 kW (huge for the time, but about average for wind turbines going into American soil today). Tinkerman talks in slow, deliberate phrases, breathing heavily as the vowels and consonants are consumed by winds as I stand with him in the Altamont Pass. "The reason I brought Howden into the United States was because they had a track record with utilities. That was important. We weren't just another start-up company composed of guys who used to sell backwater shopping centers," he explained, hinting at the legacy of a type of farming that tries to balance the rugged forces of nature—the wind's ability to destroy—and the equally brutal world of high finance. His joint venture partner, Howden, for example, also constructed cooling towers for nuclear reactors. Tinkerman sees himself as one of the many wind farm pioneers who cut the deals necessary to foster real solutions to a still resident energy crisis, even if these deals involved companies or individuals whose cultures and values were at odds with his counterculture ideals.

Tinkerman boasted that Goldman Sachs financed his 25 MW, state-of-the-art, utility-scale Howden wind farm in the Altamont Pass after passing over 33 other renewable energy projects. He noted that he did much of his lobbying for wind power on his own dime. "I've always been the outcast. And I've always had to compete against guys who had the financial resources to fly all over the country." Tinkerman saw from the beginning that the business of wind farming needed to involve big, credible corporations. Yet Tinkerman has his values, too. He turned down one job at Fayette that could have made him a millionaire "when a million bucks still meant something." Instead, he netted $20,000 that year.

The particular contraptions that sprang from Tinkerman's drive to move an infant wind farming industry into the corporate mainstream are huge wind turbines painted tan to match the arid environment. Though the surrounding hills are a maelstrom of motion, Tinkerman's turbines move much slower, wind whistling as it ricochets off the sharp points of the huge blades. Tinkerman never received any revenue from the project, and his share of the company was sold to Philip Morris in 1986.

◆

Trying to make a living from harnessing the wind in the Altamont Pass and elsewhere has always been dangerous business, both physically (a few wind farmers have been killed on the job) and more importantly, economically. Despite the one and a half billion dollars that had been invested in Kenetech and its stupendous wind turbines, which were always among the most advanced designs, the company ultimately could not sustain the fragile truce wind farming represented between technology and place, between machinery—spinning fiberglass blades and aluminum and steel tower and engine parts—and the random bullets of destruction that can be the wind.

Tinkerman's first wind power pals, Heronemus and Stoddard, planted the seeds of a venture that would evolve into U.S. Windpower and eventually Kenetech. They inspired him to drop everything to help push a technology that could change the world. From Tinkerman's perspective, U.S. Windpower was a visionary company, born from a desire to rely upon a wind turbine whose diminutive size stood in stark contrast to the humongous, out-of-scale machines that the federal government had foisted upon the general public and that would soon become junk piles.

"The story of Wind Potato is a metaphor for the whole wind industry," Tinkerman told me later "They committed what the legal system calls 'fraud.' Well, lying is the way to do business in America. If you get caught, it is fraud. But while the fall of Wind Potato is totally their fault, they had to play in a fraudulent economic environment, the world of skewed energy pricing and values," he said. Tinkerman estimated that the real price of conventionally generated power was 15 to 20 percent higher than today's posted market prices for electricity. "The real environmental costs are hidden," he explained. If the world's power markets operated on the basis of these real—and higher—energy costs, then companies like Wind Potato would not have had to overextend themselves so ridiculously and engage in such desperate antics to stay in business. Tinkerman then went off on a tangent about what he called "the power elite." "Do you think Westinghouse may have had a conflict of interest when they also just happen to be the one of the largest purveyors of nukes around the globe? The bulk of taxpayer dollars earmarked for wind power in the 1970s went to Westinghouse and General Electric, companies that also sold nuclear reactors into the international market, and companies like Lockheed and Boeing. That's what I mean by the power elite." All of the millions of dollars in taxpayer contributions to these familiar, giant engineering companies failed to produce a single successful wind turbine design, he was quick to add.

"It is the electric utilities and the oil companies that are working to keep the energy marketplace a sham," Tinkerman insisted, still sounding alarms about a plot he first rang over two decades ago. From the purview of Tinkerman, there is no real energy market because the prices of fossil fuels are so out of whack with the real costs they impose on the environment and society. "The other reason Kenetech is a metaphor for the industry," he continued, picking up where he left off, "is that it was an attempt to create a company out of some of the first visionaries. They were the most well-known company, the only real vertically integrated American company. They will tell you it was circumstances that led to their downfall. I will tell you the real story. . . ." But his voice trailed off before he finished.

Tinkerman ended up telling me only part of the story—his part. And although Tinkerman's feelings about Kenetech, or Wind Potato, are far from unique, one thing you learn quickly when talking to wind farmers like Tinkerman is that you never get the whole story from any one person. And each separate version of the same identical event never seems to quite match up with the others. No other segment of the electricity business is filled with such unorthodox characters so guilty of hyperbole and appealing promises. Tall tales, and extremely high and extremely low moments, are what the wind industry has been all about. It is not hard to imagine that in the audacity and unreliability of its would-be captors, one gains a glimpse of the wind itself, that the wind itself has created these characters.

Chapter 2

Breath of the Gods

> The wind was the cause of it all . . . human beings were involved, but the wind was the primal force, and but for it the whole series of events would not have happened.
>
> —Dorothy Scarborough

Different people have put different faces on the wind, a force that makes waves wave, trees sway, and leaves sing, but can also rip the roofs off houses and level entire cities. In *Heaven's Breath,* Lyall Watson claims there are more than 400 names for winds. The Arabic word for wind, *ruh,* also means "spirit" or "breath." The Greek *pneuma* and the Latin *animus* are words for wind suggesting a substance that is the very stuff of the soul. "The direct offspring of the earth," is the way Watson himself describes winds. "They are, however, no ordinary offspring, but wayward somewhat rowdy progeny over which there is little or no possibility of control." Watson notes that in all languages there is "the same blurring of boundaries, roots and meaning between the words for wind, spirit, breath and soul—as though each felt they clustered about the same essential mystery, and were loath to get too close for fear of startling or trampling upon it."

It is the Navajo tribe whose spiritual life is most clearly linked to the wind. In the words of one Navajo elder, "the wind was creation's first food." In their equivalent to the biblical Book of Genesis, the Navajo—a name whose translation is literally "Earth Surface People"—describe how tribes wandered aimlessly in a vast underworld, devoid of any knowledge or even ability to speak, until they encountered Wind. Appearing in human form, Wind spoke to the tribal members by climbing into their ears, teaching each of them to communicate with the others. Wind became each tribe member's conscience and guide. Over time, Wind shared knowledge about the ways of a world that lay above them, on the surface. Several Holy People eventually climbed to the earth's surface with Wind, who emerged in the form of four cardinal breezes (two male and two female), each having a different direction. These four breezes came to be recognized as the breath of the sacred mountains that border the reach of the Navajo cosmos.

Wind is fundamental to the existence of each Navajo. A member of the tribe comes to life when a "small wind" prompts a fetus to move, allowing it to breathe. Movements in the womb are evidence to the mother that the small wind has arrived and taken its rightful place. A "wind soul" also blows into one's body at birth and becomes the source of thought and action. The wind soul is independent of the body it occupies and is created by two winds, one from the father, one from the mother. At death, the wind soul returns to the Dawn Woman to inform her of how the body carried out the plan the wind soul had prepared. One translation of the Navajo word for wind soul, or knowledge, is "the silent one" or white wind, which is believed to come from the east, from the day's dawn, the beginning. There are five kinds of small winds (yellow, blue, black, white, and speckled), and each enters the child through one of five fingers. Winds are holy beings that whisper comforting advice and determine the fate of each Navajo. If a tribe member ignores the advice of the wind soul, it may leave and be replaced by evil spirit winds that lodge themselves between the eyes.

Whereas the Navajo possess an incredibly sophisticated relationship with wind, many other indigenous peoples have also long held the wind in high regard and believe that the wind is a voice of spirits. So-called "Four Winds" medicine men, a designation within different tribes of the highest degree of spiritual attainment, speak of how the directions of wind symbolized the four races of humankind: yellow, red, black, and white. Rituals were, and still are, performed to anticipate the gathering of these distinct races and their joining together as one society to communicate with higher beings in order to heal the earth.

Hindu creation stories, like those told by indigenous peoples in Mexico and told in Japan, involve the wind and tie one's breathing to the beginning of the world. Polynesian peoples assigned a variety of attributes to the wind, including its ability to carry legends, news, and even gossip about royalty. Wind has always been a phenomenon that provoked wonder among humans because of its whimsical nature, incredible ability to destroy, and fundamental role in so much of what nature represents to us.

According to Chinese medicine, the wind from any direction represents the energy of spring, the time of the year when new seeds are planted. When diagnosing a patient, practitioners associate wind with anger, one of five fundamental emotions the Chinese attribute to natural elements. During spring—a time when everything seems possible—the wind represents the anger of the pioneer, who has both the vision and the plan to accomplish his goal, no matter what.

In Greek mythology, winds were controlled by Poseidon, ruler of the great seas, and Aeolus, who tried to keep winds locked up in a magnificent whistling cavern and who was able to unleash withering blasts from his collection of tempests hidden in his long, thin sack. Eos, goddess of the dawn, was the mother of winds named Boreas, Zephyr, Notus, and Argestes.

When I was a kid, my father always told me: "Wind from the west, fishing is best; wind from the east, fishing is least." Apparently, western winds bode well for most endeavors—not just hooking walleyes in the brisk Wisconsin spring or generating electricity. (California's wind farms are designed to take advantage of prevailing winds from the west.) They typically meant stable weather. Easterly winds, on the other hand, have long been associated with turbulence and ill will. "God prepared a vehement east wind" (Jonah 4:8) is just one passage in the Bible referring to the purported dark personality of the east winds. But winds from other directions have also been blamed for calamity. Shakespeare attributed everything from the gout to itching to "falling evil" on cold winds from the north.

◆

Somewhere around 600 B.C., the Greek astronomer Anaximander uttered the blasphemous notion that the wind was not a supernatural and spiritual force, but instead was a natural phenomenon that could be examined scientifically. A hundred years later, Anaxagoras came up with the theory that heat caused air to rise and that it cooled as it climbed higher and higher, ultimately transforming into clouds. Pliny the Elder, a first-century Roman naturalist, had many theories about the origins of wind. "A breath that generates the universe by fluctuating to and fro in a sort of womb," was one. Gusts were either "the dry and parched breath from the earth" or were terrestrial agents, which came to life when "bodies of water breathe out a vapor that is neither condensed into mist nor solidified into clouds."

Science today tells us that wind is a form of solar energy, born of a perpetual effort by the atmosphere surrounding earth to reach some sort of equilibrium. The planet we call home spins on its axis, taking 24 hours to complete a rotation. Because of its rotation, mammoth streams of air circulate around this sphere, with some tributaries traveling as fast as 300 mph. Because the equator is closer to the sun than are the North and South poles, air near its surface is in constant flux, with surface temperatures rising and falling as areas move toward or away from the sun. The uneven heating of this sphere results in the moving currents of air we call wind. These movements complement and compete with each other on a global scale. Generally speaking, the bigger the temperature differences, the bigger the winds.

If the earth were a smooth sphere, the direction of wind flows would be fairly straightforward: Warm air at the equator would rise, making room for cold air from both poles to rush in to take its place along the equator. But since the earth is not a smooth sphere, a host of variables determine the personality of the wind at any one point down at the earth's surface. Landscapes as varied as deserts, grasslands, and forests, as well as huge bodies of water, add layers upon layers of complexity. Each of these environments absorbs and radiates heat at different rates. Since wind is affected by friction, hills, such as those in the Altamont Pass, reduce a wind to half the strength of a wind of equal force blowing across calm waters.

Landscapes and large bodies of water are not the only influences on the basic wind pattern of hot air from the tropics climbing into the upper atmosphere and cool surface winds from the two poles moving toward the equator. The turning of the earth conveys a curved stir to this wind regime, a phenomenon called the Coriolis effect. Gaspard de Coriolis, a nineteenth-century physicist, discovered that a moving air mass surrounding a rotating body—the earth—and moving at an angle to the direction of rotation, such as northward or southward, will be deflected from its original path by forces of inertia.

Coriolis also was the first to understand that points on the surface of the earth make progressively smaller revolutions, and therefore travel slower, as one gets closer to the poles. Think of an old LP record revolving on a turntable at $33^{1}/_{3}$ rpm. Any two points on that disk complete a revolution in the same amount of time, but a point near the center travels much less distance, and therefore at what could be considered a relaxed pace, compared to a spot on the outer edge of the LP, which must travel a much greater distance at a far faster clip to complete a revolution in the same amount of time. Likewise, a point on the equator travels eastward at 25,000 miles in one day, or roughly 1,000 mph, while a spot to the north at the level of the Altamont Pass travels less than 20,000 miles in the same time, or roughly only 800 mph. An air current traveling northward from the equator keeps its higher sideward velocity and so travels eastward faster than the surface it is passing over, as will an air current heading south from the equator. Conversely, as currents from the North or South Pole head toward the equator, they pass over surfaces moving eastward at gradually higher velocities, and thus these currents will tend westward.

The result is a predominantly counterclockwise circulation of streams in the northern hemisphere and clockwise circulation in the southern hemisphere. While the flow of air toward the equator creates the trade winds that were so critical to sailors throughout time, a flow also occurs in the opposite direction, carrying warm subtropical air farther north, where it encounters cold air masses. It is in the temperate zone, a term used for latitudes between 35 and 60 degrees in both southern and northern hemispheres, that the temperature can drop 50°F within a matter of a single hour. This wide divergence of temperatures is explained by the turbulence created when polar and subtropical air masses alternate as the dominant river of air. These wide belts of air in the northern and southern hemispheres are the key uncertainty in weather prediction.

The chief challenge for early sailors was how to buck these prevailing wind patterns, particularly when taking long trips, such as when Columbus sought a new route to the Indies. When explorers set sail, they typically crossed oceans by easily moving from east to west in tropical latitudes. Coming back to Europe was always the hard part. History is littered with tales of sailors waiting for weeks for counter winds as they drifted near the equator, in regions known as the doldrums or Horse Latitudes, so named because horses were often thrown overboard to conserve water.

Some of the worst natural disasters known are nurtured by these global wind patterns as they interact with the ever-moving, vast ocean. All it takes is a small surface disturbance to trigger the process if the surface waves are warmer than 80°F. Humid surface winds colliding from different directions generate a high-pressure system that draws a water funnel up into the dark furling sky. As the world turns, concentric thunderstorms ranging from 3 to 30 miles in diameter begin wrapping themselves in blankets of clouds, water, and vapor, all the while spinning into what might look like electrical coils if seen through a good telescope from a standpoint on the moon. One out of every 10 of these cyclones reaches speeds of 74 miles per hour or more. Once that velocity threshold is met, the storm is considered a tropical cyclone—the generic term for ocean wind storms. The power and magnitude of tropical cyclones surpass that of any other weather phenomenon. They are the greatest storms on earth. In the eastern North Pacific, Caribbean, and Atlantic, these storms are known as "hurricanes," a word derived from the Taino "huracan," or evil spirit. The myths have it that terrifying winds are wound together to punish people when the God of All Evil is angered. In the rest of the Pacific, these same storms are called "typhoons," a bastardization of the Chinese "ta-feng," or violent winds. In other places in the world, the more generic term cyclone is typically used.

Even under ordinary circumstances, the wind is a major physical force to be reckoned with. A typical summer afternoon thunderstorm has the energy equivalent of 13 Nagasaki atomic bombs. Hot winds that flow down steep mountain ridges can literally eat snow—instantaneously transforming the fluffy white into water vapor. A single blast of wind once flattened over 250 oil derricks in California.

The most powerful wind ever recorded in the United States was a 231 mph gust on April 12, 1934 at the top of Mount Washington in New Hampshire. In winter, hurricane-force windstorms averaging 75 mph pound this spot nearly every day. The annual average wind speed there is 40 mph. But the Altamont Pass does not disappoint when it comes to proving that the wind is a force to be reckoned with. The wind there can be a beast. It has been known to snap off the blades of wind turbines and hurl them as far as a quarter of a mile.

◆

Perhaps the most spectacular account of the sheer fury of the wind in a particular spot appears in the 1939 book *Wind, Sand and Stars*, by Antoine de Saint-Exupery. Here is one passage from a particularly hair-raising flight in which Saint-Exupery learned a new respect for the wind as it played with his tiny speck of an airplane:

> There I was, throttle wide open, facing the coast. At right angles to the coast and facing it. A lot happened in a single minute. In the first place, I had not flown out to sea. I had been spat out to sea by a monstrous cough, vomited out of my valley as from the mouth of a howitzer. When, what seemed to me instantly, I banked in the coast-line, I saw that the coast-line was a mere blur, a characterless strip of blue; and I was five miles out to sea. The mountain range stood up like a crenelated fortress against the pure sky while the cyclone crushed me down to the surface of the waters. How hard that wind was blowing I found out as soon as I tried to climb, as throttle wide open, engines running at my maximum, which was 150 miles an hour, my plane hanging 60 feet above water, I was unable to budge. When a wind like this one attacks a tropical forest it swirls through the branches like a flame, twists them into corkscrews, and uproots giant trees as if they were radishes. Here, bounding off the mountain range, it was leveling out the sea.
>
> Hanging on with all the power in my engines, face to the coast, face to that wind where each gap in the teeth of the range sent forth a stream of air like a long reptile, I felt as if I were clinging to the tip of a monstrous whip that was cracking over the sea. . . .

Interacting with global patterns of wind are regional quirks such as the air streams that gave Saint-Exupery such a hard time. Much of the sun's energy is reflected by the huge water mass that is the ocean; evaporation creates air that is cool. The air over land masses, meanwhile, heats up, expands, and becomes lighter. As this warmed air rises, the cooler air that resides over the ocean rushes in and takes its place. This is what happens at the Altamont Pass, where winds blow predominantly from west to east because the warmer air of the interior valley draws in the cooler air from the Pacific. (This is the dynamic that also spins the wind turbines at California's other wind farms.) In the winter, when temperature differences are often not as great in the Altamont Pass—and the Pacific Coast can actually be warmer than the valley because of chilly inland fog—winds can blow from east to west. Winds may also trace ridgeline tops by moving north to south when temperature differences between the two areas to the east and west are minimal and the winds are driven by the simple fact that the south is warmer than the north in the Altamont Pass.

Consider the various layers of air that come in direct contact with any wind turbine and one gets a better sense of the miracle that wind farming technology represents. First, a microscopic layer clings to the earth's surface by atomic forces. This is the only part of the world of air that is free of any wind. Immediately above this zone is the "laminary layer," a thin stream of air that glides parallel to the ground. The temperature swings at this level are quite wide on any given day, ranging from bitter cold to scalding hot. Just above this ever-moving yet relatively smooth layer of wind lies yet another layer of atmosphere, the planetary boundary layer. This is the zone we humans know best and is where most wind turbines derive their fuel. Here turbulence reigns as landscape features such as a knoll, a deep canyon, or even a large tree help compress and redirect the wind, and gusts are propagated when wind interacts with obstructions in the landscape. This layer of air breathes like a living organism, waxing and waning with the rise and fall of the sun. On a hot day, it can expand to 2,000 meters thick; on a frigid night, it can shrink to a mere 100 meters. Above the planetary boundary layer is a transition zone where normal friction effects gradually dissipate.

The winds that blow across the Altamont are part of a fairly unusual environment that promised profits but ultimately posed problems for wind farmers. The energy-intensive boundary layer is only a few hundred feet deep in the Altamont. In most of the world's top wind regimes, the wind energy sustains itself over vast stretches of land because of its phenomenal height and width. But the temperature differences here in the Altamont between the cool Pacific and hot central valley are not as great as at San Gorgonio or Tehachapi, in southern California. The Altamont wind regime makes less kinetic energy available for the harvesting of electricity at the upper echelons of the boundary layer. Adding to that difficulty is the topography of the land itself. A gradual sloping on the eastern ridges of the Altamont, where the vast majority of wind turbines are sited, dissipates the available energy over wide swaths of land. The winds thin rapidly as they move east.

Determining the best place to install an individual wind turbine demands great attention to detail. Measuring wind fuel requires that subtle nuances of the wind resource be taken into account so as to avoid "wind caves"—spots where the shape of a hill or the trees or rocks arrest the wind. The assessment of wind resources has come a long way and has evolved into a sophisticated science requiring numerous instruments and computer models. Regions are first screened according to annual average wind speeds and then measured and analyzed at the micro-level to ensure that each turbine is located in the best possible site for the prevailing seasonal and daily wind patterns.

The reason wind turbines are often placed as high up in the air as possible is simple physics. The wind's energy is proportional to the cube of its speed. In other words, a 10 mph wind carries almost twice as much energy as an 8 mph breeze. A doubling of wind speed yields an eightfold increase in energy production. Typically, the higher you go, the better the energy flow. But that's not really the case at the Altamont, where narrow bands of strong, violent winds cling closer than usual to the ground.

The wind has always wielded a double-edged sword. It plays a critical role in the production of crops such as corn and wheat, for example; smart farmers learn to plant in rows perpendicular to prevailing breezes and in specific arrangements to encourage pollination. Wind farmers in the Altamont and other locales install their turbines in much the same way to maximize the harvesting of electricity. Winds may be a critical factor for the success of both traditional farming and the generation of electricity, but it can twist shrubs and trees into strange shapes and cause far more damage to wind machines than a hundred Don Quixotes.

Chapter 3

Welcome to the Machines

> Is it fact—or have I dreamt it—that, by means of electricity, the world of matter has become a great nerve, vibrating thousands of miles in a breathless point of time.
>
> —Nathaniel Hawthorne

The buying and selling of wind is a time-honored tradition with roots back in the thirteenth century, when sailors sometimes paid witches high prices for enchanted knots said to release different types of winds when untied. Wind witches practiced their magic on islands off the coast of ancient Gaul. Legend has it that Scottish witches tied a wet rag infused with a hex around a piece of wood, which was then rapped against a stone as the following incantation was repeated three times:

> I knock this rag upon the stane
> To raise the wind in the devil's name
> It shall not lie till I please again

To practice the fine black art of conjuring up premium-quality gales, gusts, and squalls, one had to first become baptized but then renounce Christ. These prerequisites to entering the wind trade outraged the church, which, of course, strongly discouraged the practice. Still, the business survived despite efforts by some writers, including Thomas Ady of England, to discredit it in the mid-seventeenth century. Interestingly enough, Ady speculated that wind vendors likely formed secret alliances with astrologers, who could use their knowledge of stars to forecast near-term wind patterns. Yet even respected intellectuals, such as French surgeon Martin de la Martiniere, continued to give first-hand accounts of how wind wizards helped those in need of a good wind. One such local wizard in Lapland provided Martiniere with three knots. When each was untied, the sweetest winds imaginable provided immediate relief, filling his sails on what had just been calm waters above the frigid Arctic Circle.

Winds were still being sold as late as the nineteenth century. Sir Walter Scott chronicled the following meeting with a wind witch in the early 1800s at Kirkwall in the Orkneys:

> We clomb, by steep and dirty lanes, an eminence rising above the town, and commanding a fine view. An old hag lives in a wretched cabin on this height, and subsists by selling winds. Each captain of a merchantman, between jest and earnest, gives the old woman sixpence, and she boils her kettle to procure a favorable gale.

The Lapps and the Finns of Scandinavia had the most precise definition of wind sales. Vendors hawked magical knots on cords, but each could only claim command over wind from one direction—that which was blowing at his or her exact time of birth. The tradition of buying winds carried over into the New World. Sailors in Maine, for example, believed favorable winds could be procured by tossing money off the sides of their ships.

Between the fourteenth and nineteenth centuries, wind fuel provided as much as a quarter of Europe's total energy needs; the waterwheel and human and animal labor provided the balance. These estimates, by economic historians Sam Schurr and Bruce Metschert, take into account not only large fleets of sailing ships but also a growing reliance on wind power for terrestrial uses. The Middle East, which is best known today for its grip on global oil supplies, was also the setting for the earliest windmills. "There is in the world, and God alone knows it, no place where more frequent use is made of the winds," is the way two tenth-century Arab writers described a region now straddling the border of Iran and Afghanistan. Looking like merry go rounds, these crude mills evolved quickly. Initially, the vertical axis of the mill was surrounded by walls that directed winds toward the cloth sails that lay below the grindstones. Over time, however, the sails were placed above the grindstones in recognition that wind fuel typically improves with height. The use of windmills then spread to China before making its dramatic arrival in medieval Europe, where historians have credited the windmill with everything from the birth of capitalism to a new worldview of nature as a vast cosmological reservoir of forces that could be put to work to meet human needs and desires.

The windmills that Don Quixote attacked in Spain were perhaps an indirect result of the Christian Crusades to reclaim the Holy Land from Islam. More likely, they were the direct result of Muslim rule over more than half of Spain before the fifteenth century. A marriage between sailboat and mill technologies occurred as wind power's use for transport and mechanical power spread north and west toward Europe, or so believe many historians. Remnants of these early windmills can still be seen in operation today on the island of Crete, capturing warm, moist Mediterranean breezes with what look like boat sails. Europeans returning from the Crusades, which began at the end of the eleventh century, brought with them stories and artifacts from the exotic East that opened the eyes of many. Not only did they bring the art and architecture of these foreign cultures, but also new technologies, including windmills.

These windmills greatly improved the living standards of European women and men, whether rich or poor. Relying on wind, water, and animals for mechanical power transformed the life of the masses; women, for example, were no longer relegated to the dreary task of grinding grains by hand. "The chief glory of the latter Middle Ages was not its cathedrals or its epics or its scholasticism: it was the building for the first time in history of a complex civilization which rested not on the backs of sweating slaves or coolies, but primarily on non-human power," sums up historian Lynn White. Before the windmill, waterwheel, and other new technologies, only the wealthy possessed such autonomy. There was no such thing as social mobility.

The standard view is that wind technology originated with vertical-axis structures in Persia (such as the "merry-go-round" design described earlier), then evolved into horizontal-axis technology (the Cretan and "Dutch" windmills, for example) as this new source of power spread across the Mediterranean to Europe. Nevertheless, historian Edward Kealy takes a contrary view. In his book *Harvesting the Air*, he argues that windmill technology actually originated in southern England and then spread east. According to Kealy the horizontal- and vertical-axis technologies arose independently of one another. Regardless of which region first hosted windmill technologies, the first windmills in Western Europe were installed as protests against the King of England and the Pope, who vested riparian (waterway) rights in a select few. Securing the rights to waterways was critical to early efforts to generate power, since giant waterwheels were used to crush grains. Those who owned the water rights essentially had a monopoly on power supply and could charge large fees to struggling farmers and other citizens seeking mechanical power.

These water monopolies led to the introduction of the English post windmill, which, like the windmills in the Netherlands, spun on a horizontal axis. The first of these was erected in A.D. 1137. These English windmills initially employed sails and later evolved into machines that used roller and ball bearings and substituted metal for wood parts. Entrepreneurs realized that though the water rights were the property of the lords, the wind was free. "Brash independents—clever peasants, knights dissatisfied with their small fiefs, university-trained intellectuals, and women anxious to support themselves—soon built their own rival mills," notes Kealy.

The Church and King did not let these renegades farm the wind on their land without a fight. After all, more than one lord proclaimed, the fees collected by the Church went to support the struggle to save human souls. In 1191 a knight tore down a windmill he himself had erected. Having refused to pay the Church tithes from this mill, arguing all should have free access to the wind, he was threatened with having both of his feet cut off! Another windmill was demolished by a bishop for fear that the abbot of the monastery that owned the land, and nearby waterwheels used for milling, would discover that the windmill was constructed with lumber harvested on Church-owned land.

But efforts to stop the windmill industry were doomed. The authorities could not stop windmills from sprouting up like weeds throughout the windswept green hills of England. One English Lord, notes Jan DeBlieu in the book *Wind*, lamented that windmills offered "quick-witted peasants an opportunity to evade manorial regulation, act independently, and become quite prosperous." DeBlieu notes that this lord likely exaggerated the prospects of these early wind farmers. Many lords imposed a mill fee—called the Milling Soke (hence the expression to be soked, or soaked, by taxes) for all corn raised on their lands. Despite a ruling in 1391 by the Bishop of Utrecht that no one could own the wind since no one could control it, lords continued to try charging millers for the rights to use wind in some communities as late as 1789. Nevertheless, peasants in some regions circumvented these fees by taking their harvest to private mills operating outside of the lord's monopolistic manorial systems.

Among the innovations the English contributed to windmill technology was the "fantail," invented by a blacksmith. A fantail is nothing more than a small rotor placed on a skinny pole behind the mill and perpendicular to the main rotor. A rope attached to the fantail would help guide the main rotor directly into the face of the oncoming wind to avoid the type of tail winds that could destroy an entire windmill within seconds. Another challenge was designing brakes. Early braking techniques consisted of throwing sand into the rotor to increase friction and slow it down. Eventually, shutters not unlike Venetian blinds were incorporated into the sails. A particularly powerful gust would force open these shutters, thus reducing the surface area of sail responding to the kinetic energy of the wind. The rotor would slow down and avoid being ripped to shreds.

In the sixteenth century the center of windmill technology advancement shifted from England to the Low Countries of Europe, which lacked the options for hydrological power tapped by parts of Europe with higher elevations. The Netherlands, the country that popularized wind power, achieved a dramatic acceleration of windmill technology. Sails attached to mills that closely resembled waterwheels were employed to drain the marshes and lakes of the Rhine River delta. These huge mills and new dikes constructed to hold the water at bay increased the amount of tillable acreage by 40 percent. Land previously 10 feet under water was now being cultivated, greatly increasing economic development opportunities.

The Dutch began building industrial windmills, some of them used to process wood pulp to meet the enormous new demand for paper created by the invention of the printing press. By the late 1800s, some 9,000 windmills provided over 90 percent of the power used by Dutch industries. "Windmills belong to the Dutch landscape, to such an extent, that we cannot imagine this landscape without them," said Frederick Stokhuyzen in *The Dutch Windmill.* The principal advance in windmill technology in the Netherlands was vast improvement in rotors; these evolved from jib sails on wooden booms to aerodynamically designed sheet metal blades that featured a variety of shutters and flaps to control rotor speeds in extreme weather.

The Dutch windmill served a broad social purpose. It was certainly much more than a power plant. Since the large mills tended to be located in places where shelter was scarce, but often quite welcome because of the wind, many featured living room basements stocked with food for travelers. The windmills also became community bulletin boards, as the position of stopped sail blades indicated the current status of the mill or the state of the owner's heart. For example, the operator might signify a wedding by placing the uppermost sail in a virtual vertical position.

As the twentieth century loomed, an estimated 100,000 windmills were scattered throughout Europe. However, cheap coal did the wind power industry in, even in the Netherlands, a country whose very culture and economic prosperity were so clearly linked to this technology. Whereas windmills hastened the advance of the industrial revolution throughout Europe, the invention of steam-powered mills that relied upon fossil fuels, which could be burned any time and anywhere, ultimately rendered these symbols of independence and rebellion relics of a time gone by.

◆

In the late nineteenth century, America became the focal point of advances in harnessing the power of the wind, and northern California played a major role in pushing wind power technology in a new direction. Shortly after the gold rush was over, in the 1850s, the first windmills were going into the ground in California. Instead of grinding grain, most of these pumped water. Historian Jim Williams maintains that the first windmill in California that really worked was in the city of Benicia, which sits near the mouth of the Carquinez Straits, to the west of the Montezuma Hills, where a contemporary wind farm is sited. The center of California wind farming in the late 1800s, nonetheless, was the city of Stockton, which is still further east.

I pass through Stockton on my typical route to the Altamont Pass from Sacramento. There are few signs of any legacy of wind farming here. Few residents know that this growing urban sore was once an agricultural outpost known to as "the City of Windmills." Best known today as the host of the training camp of the San Francisco 49ers football team, Stockton probably used more windmills per capita than any other community in California in the early part of this century. One visitor commented in 1879 that Stockton's nickname appeared "appropriate to the traveler who approaches the town on a windy day, and at a distance sees little save a multitude of great arms revolving furiously above the trees and housetops." A few years later the same fellow noted that within a radius of a few hundred yards of Stockton's Yosemite Hotel "one can count more windmills than are to be seen . . . elsewhere in the state." Some nearby farms used as many 60 windmills to irrigate crops and provide water for ranching.

Other places in California had far superior winds, so the emergence of Stockton as a focal point of early windmill technology was apparently linked to local demand by agricultural users and the corresponding rise of a local windmill manufacturing industry. Typical was Abbot and Stowell, which produced the Relief Windmill. The firm had grown from a small blacksmith shop opened by Americus Miller Abbott in 1871 into a 20-person company occupying a 10,000 square foot building a few years later. All told, some four manufacturers and another four companies involved with related products and services could be found in the Stockton area in the mid-1880s.

A thriving industry of windmills was pumping water for crops throughout the San Joaquin Valley as early as 1858. There is documentation of ego-driven battles over wind machines in this region that date as far back as 1860. The San Joaquin Valley Fair in Stockton offered the general public a glimpse of a type of human behavior that would remain remarkably consistent with wind farmers as the technology evolved from a tool for traditional agriculture to an electricity generator in corporate wind farms.

At the 1860 fair that first year, two manufacturers hailing from Benicia and competing for a fair prize for best windmill design hurled insults at one another. W.I. Tustin, who sold his mills for as high as $250, claimed his main competitor's windmill was a loser. Hyde and Storer, manufacturers of the Philips mill, which sold for as much as $350, responded by noting that the company owned the patent rights to several mills operating throughout the Benicia area. They claimed that along with water pumping, they were great for driving grist mills, tanneries, and circular saws. Despite Tustin's charge that only two of the premium Philips mills were operating and they did not run well during strong gusts, the judges awarded Hyde and Storer the top prize. Tustin then challenged Hyde and Storer to a trial contest on San Francisco's highly visible Telegraph Hill. The firm, with first place award in hand, politely declined.

For two decades, windmills went up all over the state. Hundreds of the spinning pinwheels could be seen along the west side of the Central Valley, taking advantage of the same energy flows that attracted contemporary wind farmers to the nearby Altamont Pass. These water pumpers were quite different from most European designs, which faced into, not away from, the wind. The English and Dutch windmills described earlier in the chapter ultimately required wheel or pulley systems to hold the rotors into the wind. Many American windmills instead had their rotors facing away from the wind, which obviated the need for systems to hold the rotor into the wind: The rotor could now shift from right to left in direct response to the changes in wind direction, a seemingly simple solution to the challenge of capturing the most power from the wind.

Most American windmills built in the late nineteenth century were relatively small. A typical fan measured 14 feet in diameter. But a few larger mills, with fan diameters as large as 30 feet, were also constructed. Placed on 20- to 30-foot towers, these mills were nicknamed "Italian mills" because Italian truck gardeners used them extensively. Because these huge gadgets lacked weather vanes and faced into the wind like European models, they had to be tied into the wind with ropes. Though powerful, these mills were also fragile and had to be dismantled to protect them from winter storms. Taking these structures down became a fall harvest party, as friends and neighbors from throughout the Stockton area would ease the huge fans to the ground. Much wine was shared as groups traveled from neighbor to neighbor until all were bedded down. The same ceremony was repeated in the spring, only this time the mills would be lifted by block and tackle. Of course, similar levels of wine consumption were required for this important task.

Between 1860 and 1880, the California windmill industry boomed, with many observers recording windmills in the San Francisco Bay Area counties of San Mateo, Alameda, and Santa Clara, and in Los Angeles County. And though Stockton was the City of Windmills, *Scientific American* magazine recognized the San Francisco area as a great wind farming spot:

> Windmills are becoming great institutions in San Francisco. They are being extensively employed for pumping of water, propelling the shaft of the machine shop, turning the burr-stones of the flouring mill, etc. The weather here is particularly adapted to the windmill business, a large supply of wind being constantly in the market and obtainable without money and without price.

Even Sacramento was known for wind farming. In 1859, Jacob Dickerson's mill took first prize in the Sacramento State Agricultural Fair. Dickerson later boasted, by the way, that over a hundred of his locally manufactured windmills were still working a year or more after being installed. Apparently they were among the better designs, as published reports spoke highly of his windmill—other designs had "not come up to the standard of excellence desired."

At the time, close to a dozen manufacturers were located in northern California, the majority in Stockton and San Francisco. The first mills were wood, but competition from manufacturers back east, rampant deforestation, and the desire to extend the lives of the windmills led to the manufacture of steel mills in California in the early 1890s. Nevertheless, mail order houses such as Sears Roebuck and Montgomery Ward enabled out-of-state firms to continue to offer cheaper alternatives than in-state producers. The Aeromotor Company of Chicago bragged as early as 1892 that it would sell 60,000 of its galvanized steel mills that year, after recording sales of over 20,000 the previous one. By 1905, only four California firms, employing 19 workers, remained in business, two of them located in San Jose. By 1914, only one firm endured. And while Stockton became known for its windmills in California, it was Chicago, the "windy city," that became a U.S. center for windmill manufacturers throughout the latter half of the 1800s.

By 1889, an estimated 77 windmill factories were scattered throughout the country, employing over 1,100 workers and with sales in excess of $4 million. The three prime markets were railroads, which used the mills to fill their track-side water tanks; farmers, who used the water pumpers for irrigation and stock watering; and wealthy families who wanted running water for their bathrooms. The main advantage of the American windmill was that it was cheap to produce and required so little maintenance compared to the far more massive European windmills. The American windmill also became a mesmerizing element in the landscape, a symbol of self-reliance and mystery, claims writer Michael Marlone:

> It is not only that the windmill is a vertical structure in a horizontal setting which makes it such a powerful and useful image. Nor is it simply its kinetic nature, though that is a part of it. The emotional punch to watching a steam locomotive is the animation of the driving rods and pistons, the blowing smoke and steam. It moves as it moves. I've seen people weep at the sight of it. So too does the working windmill wind and pivot. Standing out above the oasis of the farmstead, the windmill blades flash and rumble. From a distance, with the tower invisible, the spinning disk seems to provide its own life, its own gravity, a tiny silver satellite at the zenith above the barn. The movement catches the eye. This country is not only wide but still. . . . The only thing moving is the wind. It is seen only in its effect as it turns the blades. The windmill moving makes the invisible visible.

The wind industry began faltering in California at the turn of the twentieth century because of increased competition from steam, gasoline, and electric pumps. These alternatives were cheap and offered the customer the convenience of pumping at any time he or she wanted, not just when the wind was blowing. The wind power industry did remain robust in the Great Plains.

Until the industrial revolution, windmills ranked second only to wood fuel as a source of power. In the United States in the mid-1800s, roughly 25 percent of the nation's nontransportation energy was supplied by the windmill. A number of wind pumpers continue to work today near the Altamont Pass, Montezuma Hills, Stockton, and locations throughout the country, but most particularly in the Great Plains and midwest.

Chapter 4

Tesla's AC and Edison's Utility Monopoly

> Of all the forces of nature, I should think the wind contains the greatest amount of power.
>
> —Abraham Lincoln

Many attracted by the allure of mechanical power and steam engines during the latter half of the nineteenth century thought windmills were crude relics of the past. But one famous inventor, Nikola Tesla, endorsed wind power around the turn of the twentieth century. The endorsement didn't necessarily lend much credibility to the wind power industry, however, since Tesla's sanity was debatable.

Born in Croatia at about the time that America began tinkering with windmills, in the late 1850s, the tall and lanky Tesla was a handsome and dashing Serb who popularized the image of the "mad scientist" as he repeatedly subjected his body, as well as a variety of animals, to immense electric shocks. His late night experiments, sometimes in the company of buddy Mark Twain, became legendary in the annals of the bizarre. Twain loved Tesla's electrifying acts. "Let's have the show," he once exclaimed after meeting him at his West Broadway laboratory in Manhattan, at the stroke of midnight. "Thunder is good, thunder is impressive, but it is lightning that does the work," Twain would often say, referring to Tesla's highly visible displays of the power of electricity.

Jumping on a platform that would subject him to thousands and thousands of volts of electricity, Tesla was indeed a show-off. The intense high-voltage electricity would shoot out of every inch of his body like flames, his tall frame a mere silhouette of a man. His displays of what extreme electric currents could do were mostly done for publicity—but with a purpose: He was proving that high-voltage alternating current (AC), largely flowed on the outer surfaces of the skin and therefore did no harm (though he warned that sustained exposure could damage nerves and be fatal). Tesla lived to be 87 and made a number of important discoveries. His first patent, granted in February 1886, was for an electric arc lamp. Although much of his lab work addressed the practical implications of widespread reliance on electricity, his "inventions" became more controversial over time. In the twilight of his career, Tesla garnered immense notoriety for an ultrasound gun, which he dubbed his "Death Ray." It inspired a 1938 horror flick starring Boris Karloff and made its way into the very first Superman cartoon, entitled *The Mad Scientist,* in the 1940s.

Tesla, the so-called "New Wizard of the West," endorsed wind power in 1902, believing that it should be developed in order to conserve finite reserves of biomass and fossil fuels. Windmills, he argued, could be placed on the roofs of houses and used not only to pump water but to power elevators and cool houses in summer and heat them in winter. Lord Kelvin of England was a big fan of Tesla. Inventor of the Kelvin scale, which measured "absolute zero," a temperature so low that molecules cease motion, he announced he was in complete agreement with Tesla on the issue of wind power. Lord Kelvin was particularly concerned that Britain's dependence upon coal was decimating forests. To gain access to coal deposits, trees were clear-cut and huge mines were developed in regions that were once undisturbed, wooded landscapes. Kelvin believed that since fossil fuels were in finite supply, costs would inevitably go up. It was not "utterly chimerical to think of wind superseding coal in some places for a very important part of its present duty—that of giving light," said Kelvin.

But Tesla and Lord Kelvin agreed on another topic, too: The planet Mars was sending signals to earth. It was speculated by members of the press at the time that Martians might have been guiding Tesla's remarkable work, particularly his examination of wireless world communication systems. Of course, Tesla had for quite some time bragged that life on Mars was a "statistical certainty" and that he planned to communicate with Martians via wireless signals that could also transmit electric power over great distances.

Thomas Edison, already well on his way to becoming an international hero, discounted such nonsense. And on the topic of wind power, he proclaimed there were ample conventional fuel supplies to last over 50,000 years.

Indirectly, Edison first attracted Tesla to America in 1884. Tesla had learned quite a bit about electricity while tinkering in France and Germany with several early, experimental power plants designed by Edison. Charles Batchelor, a close friend and assistant to Edison, was impressed by Tesla's ability to quickly fix many of these early generators. He urged Tesla to go to America, where there were greater opportunities to strike it rich. Tired of being broke all the time, Tesla followed Batchelor's advice and arrived in the U.S. with a few pennies and poems in his pocket and a letter from Batchelor urging Edison to hire him. Tesla immediately went to visit Edison, and as luck would have it, Edison was desperate to find someone to fix a number of power plants in New York that were short-circuiting, and sometimes starting embarrassing fires.

Edison was initially quite taken with Tesla, but their diametrically opposed methods of work and values soon doomed the relationship. Although Edison is credited with the invention of the light bulb and dozens of other electrical devices, a few, including Tesla, have questioned the extent of his actual contributions to engineering and design. Critics go so far as to say that while Edison was a prominent inventor, his true claim to fame is as a shrewd businessman who profited dearly from the work of others. Since he received credit for all discoveries made at his laboratory in Menlo Park, New Jersey, where several other talented engineers and inventors labored, it is difficult to confirm or deny these allegations. Products such as the light bulb and other noteworthy successes were more often than not the results of work among an array of Edison's associates.

Tesla was the inverse of Edison. He did most of his inventing in strict solitude. Once Tesla came up with something, he was eager to show and tell. Tesla was, more importantly, a utopian who believed electricity and a variety of his inventions would provide limitless light and energy, end world hunger, and link human civilizations with those on other planets. Whereas Edison excelled in the world of patents and compensation, Tesla was always begging for large sums of money from Wall Street barons such as J.P. Morgan, Edison's initial financial partner, at stylish New York City parties. Tesla was cultured, mysterious, and a fanatic about personal hygiene, requiring 19 pristine napkins at his side when dining; Edison was a man of simple tastes who seemed to have an aversion to bathtubs.

The most fundamental difference, nevertheless, was that Edison always focused on the immediate and practical, whereas Tesla's vision was clouded by his extraordinary love of high-voltage power. To Tesla, electricity was a mystical force that could be transmitted without wires throughout the globe, and should essentially be community property. In light of this egalitarian view, some called Tesla the "Father of Free Energy."

The business relationship between the two ended after Tesla had worked for a year improving the design and performance of Edison's power plants. Tesla's understanding was that Edison had offered him a $50,000 bonus if he was successful, but according to Tesla, when he asked for the payment Edison retorted: "You don't understand our American humor." (Edison's version was that Tesla offered to sell him his AC patents for $50,000, and Edison thought he was joking.)

This was no laughing matter for Tesla. He stormed out of Edison's office and never returned. No love was lost. After all, knowing Tesla's gypsy heritage and the proximity of his home to what was reputed to be vampire country, Edison once asked Tesla, in all sincerity, if he was fond of meals of human flesh! Tesla soon became Edison's chief rival: "If Edison had to find a needle in a haystack, he would proceed at once with the diligence of the bee to examine every straw after straw until he found the object of his search. I was a sorry witness of such doings, knowing that a little theory and calculation would have saved him 90 percent of his labor," he once said.

Tesla established the Tesla Electric Company in 1887 in New York City with the express purpose of patenting his one-, two-, and three-phase AC generation, transmission, and distribution systems. At the time, no one had yet come up with a decent AC motor. Within the first six months, Tesla filed his first of 40 patents involving AC. For each of his AC phase systems, Tesla designed the necessary dynamos, motors, transformers, and automatic controls required to serve end-users. George Westinghouse immediately saw the immense value of Tesla's inventions, and offered the struggling inventor some cash, stock, and royalties on future AC electricity sales. Tesla accepted. His total compensation was roughly $60,000. At last he was no longer broke.

Westinghouse was one of the first to recognize how long-distance transmission of electricity by way of Tesla's AC systems represented a prime business opportunity. Yet Edison still firmly believed in direct current (DC) power systems, and was preoccupied with figuring out the mechanics of a DC-based electric monopoly system. Since electric motors, heaters, lights, and most other devices all run on DC power, Edison's position was not too surprising.

The great rivalry between Tesla and Edison centered on which was better: AC or DC. Electric generators produce DC power: current that flows in one direction and one direction only. AC is more complex than DC. AC current flows first in one direction and then the other. It is one-, two-, or three-phase current according to the characteristic variations of the electric charge (which are usually described by sine waves). Transporting electricity over great distances requires high voltages, and these can be achieved only by transformers that work only with AC.

Still, Edison engaged in a vicious campaign to discredit Tesla's array of AC-related inventions, calling AC dangerous and adding, "we are set up for DC in America. People like it." Edison feared high-voltage electricity. The low-voltage DC allowed for the proper operation of his beloved light bulb and reduced potentially dangerous sparking. But since DC current traveled only a few short blocks, his DC dynamos had to be scattered throughout a city in order to provide steady and reliable electricity. Edison embarked on a number of highly public electrocutions of animals with AC (mimicking some of Tesla's controversial research techniques) and lobbied state and local officials to outlaw high-voltage AC. It was to no avail. History sided with Tesla—but only on the issue of AC transmission.

◆

The late 1800s was a time when new technological advances in the production and consumption of electricity began to quickly emerge from the laboratories of both Tesla and Edison. This new "substance," and all of the gizmos and useful appliances it powered, fascinated much of the general public. Electricity was heralded as an invention that reflected progress and society's growing enlightenment. It could drastically alter lifestyles and was equated with a more civilized world, one that was no longer governed by nature's cycles of day and night. Electricity soon became associated in the minds of consumers with the wizardly services it provided.

Other inventors entered the picture, too. Charles Brush, a Cleveland inventor and entrepreneur, powered twelve electric arc lamps of his own design in downtown Cleveland with a single dynamo in 1879. By the end of 1880, Brush's arc lamps were lighting up Broadway, New York's "Great White Way." Meanwhile, Edison's incandescent light bulb had yet to make a commercial appearance. A year later, the Brush Electric Company (a forerunner of General Electric) tried to finagle an exclusive franchise for lighting services from the city council in Wabash, Indiana, but local leaders turned him down. Nonetheless, the concept of exclusive service monopolies from a central electricity source was born. An intense debate soon began to rage over the virtues of having this miraculous substance, electricity, supplied by private enterprise or local community-controlled public agencies. Edison modeled his electricity distribution system on the gas power distribution systems that had been in place for a couple of decades. He even assumed he would use existing gas pipes to carry his electric wires throughout buildings and would place his electric lamps in converted gas fixtures. Customers would be charged per unit of consumption.

Edison, with the backing of investment banking tycoon J.P. Morgan, had built New York City's first electric generating plant in 1882. But nobody then viewed electricity as a great money-making enterprise. It wasn't until municipal trolley systems began to be built between 1887 and 1895 that the electric transmission lines that had been idle during the day (greatly reducing potential revenues) began to operate nearly round-the-clock. With trolleys and lighting fully utilizing the capacity available on power lines, Edison began implementing a business strategy whose ultimate success would render an economic and environmental legacy far surpassing that of any of his individual inventions.

Edison's protege Samuel Insull developed the bright idea of avoiding competition in the capital-intensive electricity business by carving out exclusive, privately run franchises. These eventually allowed monopoly providers of electricity to capture huge, and profitable, markets. Insull proposed that states rather than localities regulate electric service. Though this policy was painted as being in the public interest, it, along with Tesla's AC, enabled the utility monopolies created by Edison to succeed far beyond anyone's expectations. Utility companies secured large pools of captive customers, since Tesla's AC allowed them to connect power plants over vast areas, limited only by statute and state lines. The centralization of generating capacity made economical ever-larger power plants.

This led to the massive electric grids of today. The key to it all was Edison's belief that electricity should be viewed as a commodity. It needed to be calibrated and sold in units. Many utilities on both sides of the country—Southern California Edison and Commonwealth Edison—bear his name in honor. Indeed, Edison amassed a large fortune at his death, though most of it was derived from his long list of inventions.

Tesla, on the other hand, continued to struggle financially and shifted his efforts to developing "wireless transmission" systems, which he claimed would allow power plants to distribute power anywhere in the world. While his credibility as a scientist may have been lost in the twilight of his career, Tesla did have a lasting impact on the future of electricity and ultimately the evolution of the wind power industry. Though Tesla advocated wind power just after the turn of the nineteenth century, his work on high-voltage AC motors and transformers, ironically, helped kill the renegade use of wind-generated DC power in regions not served by the utility monopoly AC grid.

◆

In a curious twist, Charles Brush, who in 1880 in Wabash, Indiana, first attempted to monopolize the market for electricity, was also the man who generated the first significant amount of electricity from the wind. In 1888 he hooked up a mill with a 60-foot tower. It featured 144 blades on a rotor that was 56 feet in diameter and covered 1,800 square feet in total blade surface area. Brush built the 12 kW, DC machine because of the high transmission costs of bringing power to the sparsely populated prairie—a major impediment to the expansion of electricity. Though the utility companies inspired by Edison were indeed in the process of securing monopolies in the eastern United States, DC power still predominated in rural regions of the Midwest. Heralded as the world's first wind power plant, Brush's invention powered his residence and laboratory. His mill looked like a cross between the stout European mills and the multi-bladed water pumpers that had become so popular throughout the Great Plains and the West.

The trend toward urban-based, centrally located dynamos, whose output was distributed throughout neighborhoods and whole towns, encouraged by the likes of Edison, ran directly against most wind farmers' cultural values and vision of the future. Those who had relied on the wind to pump water and provide other necessary services had learned to live with the rhythms of nature and prided themselves on their ability to survive without help from city slickers. Many of these wind farmers followed in Brush's footsteps and were generating

DC power on a small scale from the wind. The transition from harnessing wind for pumping water to using it for driving an electric generator was not the result of public policies or government plan, and it did not happen overnight. It was do-it-yourselfers, tinkering in their backyards and slapping together what they often viewed as giant hobby toys, who started this revolution in power-generating technologies. A few became successful wind energy entrepreneurs, the most noteworthy being Marcellus Jacobs, who came up with a user-friendly wind turbine model that was the unrivaled mainstay for "off-the-grid" rebels in the United States from the 1920s until the 1960s.

Jacobs was the quintessential renegade wind farmer, hell-bent on generating a future in which one could be energy self-sufficient with nothing more than the wind as one's primary power source. Jacobs was born in North Dakota and raised in Montana, two states that are blessed (or cursed, depending on one's perspective) with incredibly persistent, often screaming winds. He had little formal engineering training, but he used it and an intuitive sense of good design and practicality to secure more than a dozen patents on what came to be the world's state-of-the-art small wind turbine. The first Jacobs wind turbine, in 1922, consisted of a rear axle from a Model T and the fan blade assembly from an old water pumper. It worked, but there was much room for improvement.

In reworking his design, Jacobs elected to reduce the number of rotor blades from more than a dozen to three. For these he adapted propeller blades, changing the in pitch to better capture the kinetic energy of wind. Energy production increased by a factor of three. He made numerous other incremental advances and paid close attention to the fine details of the manufacturing process. Jacobs actually would fly to the West Coast to hand pick wood from a special variety of spruce that he then had crafted into rotor blades, shunning the metals and composite materials favored by most other blade manufacturers. Ultimately, he moved from the Big Sky Country to Minneapolis, Minnesota, and opened a factory to mass-produce 1 and 2 kW DC machines. In his heyday, he employed up to 260 people and produced more than $50 million worth of wind turbine machinery before closing his doors in the 1950s.

How reliable was the Jacobs turbine? U.S. Navy Admiral Richard Byrd took a Jacobs wind generator to the South Pole and left it in place when he broke camp. Twenty-two years later, visitors found it still spinning and producing power despite two decades of steady operation in the harshest of environments.

The rise of private electric monopolies that refused to serve sparsely populated rural regions, which increasingly relied on wind turbines like those manufactured by Jacobs, created a political opportunity—and Franklin D. Roosevelt responded aggressively with the Rural Electrification Cooperative Program. The public power movement, an outcome of his New Deal, encouraged rural cooperatives and local governments to create their own exclusive franchises to sell consumers the cheapest power in the land. To spur economic development,

FDR authorized construction of huge federally subsidized hydroelectric power plants. Municipal utilities, rural cooperatives, and other public power entities were given first dibs on this heavily subsidized hydroelectric power. One by-product of this hydroelectric construction was a massive effort to expand the electric grid throughout the Great Plains and points west. Some wind farmers were up in arms about this effort to link everybody in the country to one great electric grid. A few occasionally threatened to blow up the massive transmission towers that would traverse their properties.

To most rural folks, however, a steady source of electricity was quite appealing. Wind-driven electricity generators typically produce DC power because of the mechanics of converting kinetic energy to electricity. Converting DC to AC requires a synchronous generator. Farmers were now offered the convenience of AC power generated from huge subsidized hydroelectric plants and delivered right to their homes, day and night, without any need for converters, special storage batteries, or any other gizmos.

Some held out against the electric grid for quite some time. One was Joseph Spinhirine, a wind enthusiast residing in the Texas Panhandle. While all of his neighbors were signing up to start receiving power from the large hydro dams, Spinhirine held out and in 1947 purchased a new Jacobs for under $100. Not only did he still pump water with a traditional Aeromotor windmill, but he also enjoyed the comfort of a wind-electric window air conditioner. "All it takes is a lot of stubbornness and a lack of money!" was the way he summed up going solo with wind power. But folks like Spinhirine were rare.

By the 1950s, it became cheaper to buy electricity from monopoly grids that evolved from the bright ideas of Thomas Edison (and the huge subsidies afforded hydropower) than to farm the free wind. The public power movement all but killed America's infant wind farming industry.

Chapter 5

The Winds of War

Every great advance in science has issued from a new audacity of imagination.

—John Dewey

War has always inspired technological progress. Over the centuries, innovations in harnessing the power of the wind propelled a wide range of weapon-wielding ships. After World War I, coal and other fossil fuels were abundant. Cheap subsidized hydroelectricity transported over the ever-expanding electric grid. The windmill industry that had grown up to serve self-reliant wind farmers in the heartland of the United States suffered. But fossil fuel shortages and rationing during the Second World War spurred dramatic breakthroughs in wind turbine technology, as did newly acquired knowledge about the wind derived from the growing use of sophisticated aircraft.

After World War I, under the terms of the Treaty of Versailles, the Germans were denied the use of powered aircraft. This prompted German inventors to turn to gliders. In 1921, in the Rhön Mountains, Wolfgang Klemperer rode clear-air thermals to successfully complete a flight of 13 minutes and three seconds. Within a year, flights of two and three hours were quite normal. Most gliders today are based on German designs that evolved from the work of Klemperer and his German cohorts. The body of knowledge gathered in the development of these gliders would have a profound influence on German wind turbine designs.

Ulrich Hutter was among Germany's top glider experts during World War II. His initial interest in wind power stemmed from Adolf Hitler's search for alternatives to scarce fossil fuels. As chief designer for Ventimoter, a Nazi-subsidized business created to develop alternative power sources, Hutter designed wind turbines that strongly reflected his previous work with gliders. His designs, which appeared in blueprints toward the end of the war, show the German fascination with lightweight materials; they were sleek, extremely lightweight, and very, very futuristic. But the most distinguishing feature of the 100-kW wind turbines was lightweight, constant-speed rotors that were controlled by variable-pitch propeller blades. The blades in particular were revolutionary, being made of composite carbon-epoxy or composite fiberglass-epoxy. The other key design feature was a teetering hub that enabled the rotor to respond freely to changes in wind direction by teeter-tottering and swiveling in response to gusts and gales, and the like. Hutter had more influence than any previous inventor on U.S. wind turbine designs and their bias toward light, downwind machines. Hutter's work with composite materials helped advance the knowledge base that is at the core of much of today's work to reduce mass and weight. The one critical aspect of Hutter's work that American designers seemed to ignore was the catastrophic failures.

Though Hutter made significant advances in the state of knowledge of wind power, he did not construct any actual machines until well after the war. However, the Nazi occupation of Denmark would turn out to be fortuitous for the wind power industry. As the war dragged on and scarce fossil fuel supplies were routed to the military, the Danes frequently were without power. The lack of a dependable electric grid prompted the enterprising Danes to erect wind-driven electric generators, building on a little-recognized Danish legacy of ingenious advances in wind power.

Back in 1891, a Danish high school teacher named Poul la Cour had started experimenting with the idea of capturing the power of the wind. Denmark lacked domestic fossil fuel supplies. As a consequence, in the 1890s, while small water-pumping windmills dotted large stretches of America, particularly across the expansive prairies of the Great Plains, the Danish government was installing large wind systems throughout the Danish countryside. By the turn of the twentieth century, some 2,500 windmills supplied roughly 25 percent of Denmark's industrial power—the equivalent of 30 MW. This was the highest level of dependence on wind power of any country in the world. Over 90 percent of these large industrial windmills were located in rural areas, and another 4,600 smaller windmills were used on farms for threshing, milling grains, and pumping water. By 1908, more than 70 of la Cour's wind-driven electric generating machines, ranging in size from 5 to 25 kW, were harvesting the Baltic Sea winds battering the Jutland peninsula. To put these machines in perspective, the largest of la Cour's mills was comparable in size to the initial wind turbines installed in California in 1980. These windmills helped keep Denmark's economy going during World War I, when its ports were blockaded.

Denmark has a strong tradition of family ownership of farms. Most of these farmers also belong to cooperatives. The combination of self-reliance and community cooperation has been critical to the evolution of the Danish wind industry. Agricultural cooperatives and individual farmers own most of the wind turbines still spinning in Denmark today. But even in Denmark, interest in wind power faded with the end of World War II. Without the fear of blockades, Denmark resumed its fossil fuel imports. By 1944, some 88 large wind turbines were generating electricity in Denmark. By the 1950s, despite this country's legacy of innovation in wind farming, wind power was viewed as an experiment rather than as a commercial enterprise.

One student in the school of wind turbine electricians that la Cour had started was Johannes Juul, who began working on what would become the Danish standard design during World War II. It was a three-bladed, upwind machine—instead of the two-bladed downwind design favored by the Germans and Americans up until this point. Juul was the prototypical Danish wind farmer. Although he had received very little advanced education, he had been a farm equipment craftsman with an eye for good engineering fundamentals. His Gedser mill, rated at 200 kW, was not the most attractive piece of machinery in the world, but it worked remarkably well. Juul was greatly influenced by designs used by F.L. Smidth, a cement company that had installed 60 wind turbines in the 1940s. Influenced by these contemporary designs, the Gedser mill had a concrete tower and featured crude blades, only the tips of which could change pitch in response to changes in wind velocity. Though it was by far the simplest of designs to emerge during World War II, the Gedser mill boasted the best performance.

Despite the tremendous advances made in Germany and Denmark leading up to and during World War II, the first truly large, utility-scale wind turbine would go in the ground in the United States. As was the case in Europe, a primary driver of these fresh advances in wind power technology was fear of war-induced fossil fuel shortages, but another reason was the notion that the wind was free. Harnessing electricity from a substance that seemed to be abundant and that would otherwise go untapped proved to be enough of a lure for enterprising Americans to push wind power technology to the next level of sophistication.

Palmer Putnam was a Boston engineer who was best known for his invention of a wheeled amphibious vehicle known as the DUKW. Having developed 10 new weapons during World War II and directed the development of over 20 more, Putnam was no stranger to engineering science; he was awarded the Medal of Merit by Franklin D. Roosevelt for his many contributions to the war effort.

His initial interest in wind power was driven by his desire to power his own residence with a free fuel. Putnam wrote:

> In 1934, I had built a house on Cape Cod and had found both the winds and electric rates surprisingly high. It occurred to me that a windmill to generate alternating current might reduce the power bill, provided the power company would maintain standby service when the wind failed, and would also permit me to feed back into its system as dump power the excess energy generated by the windmill. But when I came to compute the size of the windmill needed to carry the peak load of the all-electrical house in the prevailing wind, it was clear that none of the small units, widely used for farm lighting, would be large enough. A much larger unit would be necessary.

Putnam discovered that there was no consensus about wind turbine designs. Studies written as far back as 1759 claimed that a horizontal-axis mill was more efficient than a vertical-axis mill. Yet some of the most recent influential work was that of Georges J.M. Darrieus of France, who performed pioneering investigations into vertical-axis wind turbines in the 1920s. Darrieus conceived several designs, the most famous of which resembled an eggbeater; it featured two long, thin, flexible blades instead of the big sails that had been the tradition with most European windmills. He thought these blades would be more durable than those of other vertical-axis configurations. Despite the advantage of being omnidirectional, Darrieus's rotors (usually aluminum) could bend or break in stiff winds, and any substantial wind imposes severe stress because it strikes the top and bottom of the vertical axis at different speeds.

Other inventors whose turbines would be classified as vertical-axis machines included Anton Flettner of Germany; his vertical wind cylinders successfully propelled a ship across the Atlantic Ocean. Flettner also built a 66-foot diameter four-armed "windwheel" each arm of which was composed of three spinning aluminum hollow cylinders. When the prototype built for the Public Service Corporation of New Jersey failed to live up to its promise, the project was canceled. Captain Sigurd N. Savonius of Finland also put forward an odd vertical-cylinder wind turbine. Though it worked, material costs were too high, making this approach an inefficient and very expensive way to generate electricity.

None of these vertical-axis designs appealed to Putnam. The design that intrigued him most was a more traditional horizontal-axis machine that had been dubbed the first successful "large" wind electricity generator of note. It started producing power on a bluff near Yalta, overlooking the Black Sea, in 1931. This pioneering, 100 kW power plant featured two blades and a 100-foot-diameter generator mounted on an inclined strut. Like some other early wind turbine designs, it moved around a circular track so that it could be positioned and secured to face directly into the wind. It generated almost 300,000 kWh annually. Despite its remarkably consistent performance, the Russians abandoned the experiment at the outset of World War II. Putnam thought that if the Russians could produce so much power with such a crude contraption, the United States could certainly do better: "I felt the principal weakness of the Russian design—low efficiency, crude regulation and yaw control, high weight per kilowatt, and induction generation—had been imposed upon the designers by the state of industry in Russia, where heavy forgings, large gears and precision instruments were unavailable."

The wind dynamo Putnam ultimately designed, with a little help from some friends, would be over 10 times the size of the Yalta turbine. It was to be built at an anonymous spot near Rutland, Vermont, which the team referred to as "Grandpa's Knob" because of its shape and the fact that the folks who sold the property would refer to the site only as "grandpa's." This is a rural part of central Vermont, where the Green Mountains are aptly named because they are thick with verdant forests. Because the surrounding ridges helped focus winds on Grandpa's Knob, it was an ideal site for a turbine. Nonetheless, this massive piece of machinery created quite a stir, its silhouette serving as a dramatic statement of a whole new dimension to wind farming. It was no longer the domain of backyard tinkerers from the Midwest. Elite engineers from New England were involved in what became heralded as a grand patriotic venture.

Putnam assembled a wide array of talent, including a General Electric Company vice president, the president of the Carnegie Institution of Washington, the former head of the meteorology department at MIT, the head of MIT's civil engineering department, and a number of high-ranking individuals from the engineering and construction firm S. Morgan Smith Company, including president Beauchamp E. Smith. All told, some 350 engineers and other experts worked on the cutting-edge machine. "This experiment is another proof that the spirit of exploration and adventure had not yet died out in those ancient citadels of capitalism, New England and Pennsylvania," declared Putnam. Vannevar Bush, dean of engineering at MIT and Putnam's boss during the war, commented in the introduction to Putnam's *Power from the Wind:*

> Complex as are the electrical systems of which it was a part, and the economic system out of which it grew, the wind-turbine is notable as the physical result of a project conceived and carried through by free enterprisers who were willing to accept the risks involved in exploring the frontiers of knowledge, in the hope of ultimate financial gain.
>
> Note that such financial gain would not have been at the cost of some other part of the economy. Large-scale wind-power will not create unemployment, for, in general, it displaces nothing, but rather draws on a new source to supplement steam and water, and can enlarge the use of the product it creates. It could, fully

> developed, bring light, heat, and power to regions that otherwise could not afford such services, or, in fact, because of physical difficulties, could not have them at all.

Central Vermont Public Service Corp. (CVPSC), a leading hydro developer based in the Green Mountains, accepted a Putnam proposal to build the wind turbine to serve as an auxiliary to the company's existing water power stations. The rationale for investing in this new source of power was that when the wind blew hard enough to generate electricity the utility could conserve its water power or other fossil power sources and store them for later use.

CVPSC became the first electric utility monopoly in the world to purchase wind power, on October 19, 1941. The machine, which faced downwind in what would become an American design tradition, stood 120 feet above the summit and featured two 8-ton blades, each measuring 70 feet across and 11 feet wide. With its polished, sunlit blades flashing, the giant Smith-Putnam machine fed wind-generated AC electricity directly into a high-voltage utility grid for the first time. A synchronous generator converted the DC current to AC. The $1.25 million machine produced 1,250,000 kW (1.25 MW) of electricity, enough for your typical small New England town.

The first major mishap occurred in February 1943, when a main bearing failed for reasons unclear. Because of the war, it took two years to find and reinstall one. The project continued over those two years as the team continued to refine and simplify the design. Then, in the dead of a cold March night in 1945, 120 mph winds shook the monster. The machine had successfully survived 115 mph winds, but the extra five miles per hour was just too much. The lone operator in the control room was knocked off his feet when one of the blades broke off. It landed on its tip 750 feet down the side of the hill.

Because the company lacked funds, and because federal support for wind power waned with the end of World War II, the machine never spun again. All that is left today are remnants of its cracked foundation.

In spite of the camaraderie the Smith-Putnam machine engendered among its long list of Yankee designers and manufacturers, it fell victim to Dwight Eisenhower's "Atoms for Peace" campaign, which redirected the imaginations of scientists at MIT and elsewhere from unpredictable wind power to the divinity of nuclear power. Though the Federal Power Commission sponsored a wind turbine design almost six times the size of Smith-Putnam, with two rotors placed on top of a sky-scraping 475-foot tower, legislation authorizing this gargantuan turbine died in 1951 with the outbreak of the Korean War. It was not until the energy crisis some 20 years later that anyone in the United States would build a wind turbine at utility scale, let alone that of the Smith-Putnam machine size.

Chapter 6

Visions of Heronemus

> There are two fools in every market: one asks too little; one asks too much.
>
> —Russian proverb

William Heronemus was a dangerous man suggesting an audacious departure from the status quo. Because of his nuclear navy credentials and impressive grasp of convincing facts and figures, his call for fleets of huge wind turbines on ships floating off America's shores in the early 1970s received considerable media attention. *Science* magazine called him "a prophet," and suddenly considerable press was devoted to the promise of renewable energy sources such as wind. Nevertheless, many in the electric utility industry and within the ranks of government worried about his radical rhetoric about the high cost of nuclear power and his endorsement of a hydrogen energy economy wherein generated electricity would be used to convert hydrogen into liquid fuel to run both cars and power plants.

A former U.S. Navy captain, Heronemus had once thought that nuclear power could become the primary power source, but he turned his interest to renewable energy when he came to believe that the real costs of nuclear power were omitted in the calculations of Admiral Hyman Rickover, his superior officer and a man he had once idolized. Heronemus had been awarded a grant to investigate the cost effectiveness and feasibility of placing submerged nuclear reactors off America's shores. He never completed the research because he soon realized that placing nuclear power plants underwater in harbors would never be cost effective. The full expense of dealing with the radioactive waste from reactors had never been truly revealed; and that number was so large that it rendered nuclear power a very expensive and irresponsible way to keep the lights on.

"I became afraid that we were creating a monster. Compared to the designs of nuclear reactors for the Navy, what the private sector was coming up with was ridiculous," said Heronemus. Still, even environmental groups such as the Sierra Club were advocating nuclear power in the early '70s. Heronemus was disgusted with the propaganda touting the construction of new utility-owned nuclear power plants and colossal breeder reactors. Proposals to line the Connecticut River with nuclear power plants alarmed him, too. "I decided that there is indeed a conspiracy. The utilities and energy companies are going to get us hooked on nuclear power, then they are going to drive the price up wherever it will go. And who are they? They are the same friendly souls who are selling us coal and petroleum products." At the time Heronemus was not so much concerned about the safety of new reactors as indignant about the false, "too cheap to meter" economic claims, which the populace soaked up like a sponge. He knew there had to be a better way.

Heronemus just happened to be passing through San Francisco shortly after World War II when he stumbled upon a book that rekindled a childhood interest. The book was *Power from the Wind,* by Palmer Putnam. The more he read about Putnam's work designing and building the world's first colossal wind turbine, the more inspired he became to devise his own schemes to address what he perceived to be one of America's great challenges: an affordable, clean, and secure energy supply.

Heronemus's relationship with wind power began on a farm, so he was very familiar with the plight of rural America and its reliance on windmills to provide much of the mechanical power on the farm. As a youngster, Heronemus watched his Uncle Henry invent a way for the wind to turn a crank to separate cream from milk at his Wisconsin dairy farm. "If you were a dairy farmer in the good old days in Wisconsin, you either sold your milk to the cheese factory or separated it and sold the cream to the local creamery," he recalled.

Born to a poor farming family that endured the embarrassing poverty of the Great Depression, Heronemus grew up to become a partisan of Senator Robert LaFollette, who had earlier revitalized the independent Progressive political party in 1924. At LaFollette's urging, he left Wisconsin for New Hampshire to serve in the U.S. Navy. After a stint as captain of a destroyer in World War II, he spent the last 10 of his 17 years under the command of Admiral Hyman Rickover, witnessing firsthand the Navy's switch of its submarine fleet from diesel electric to nuclear power. When Heronemus left the Navy behind he sought opportunities to apply his engineering skills to something other than instruments of war.

The large scale of the Putnam machine fascinated Heronemus—as did the work of Ulrich Hutter. Heronemus also respected the community values the Danes attached to wind power in their culture. Like Putnam, Heronemus was driven by a patriotic zeal. And he thought big—really big—both in scale of technology and in terms of national programs that reflected his roots. He wouldn't call himself a socialist, but his belief in big government programs was shaped no doubt by FDR's New Deal.

Heronemus made his first big splash in 1972, with a paper dispelling the dominant view that there wasn't enough wind of significant speed to generate much electricity in the United States. His numbers showed it was economical to generate electricity in places such as the Great Plains. According to his estimates, enough wind power could be installed there to generate the equivalent of 189,000 MW of nuclear power, an amount of electricity, he deadpanned, that "might properly be called significant."

Heronemus claimed that New England could obtain 20 percent of its electricity growth between 1970 and the year 2000 from the wind. Wind power, he said, "certainly can be the essential ingredient of pollution-free power systems. And it is such a gentle alternative to high temperature combustion, fission, and fusion schemes!" A lone voice in the wilderness in the 1970s, he questioned the need for any further investment in nuclear fission or fusion research. "Would there be some unacceptable stigma to our society were we to opt for an energy system whose science and technology would be very unsophisticated?" Heronemus was speaking to a widespread perception that wind power represented an inferior form of technology, certainly not as dignified as nuclear or coal plants in the eyes of utility company bosses.

However, the intermittency of the wind was viewed as its greatest drawback. Sure it was free, but you could never depend on it. Or so was the popular image of wind. But in one respect, wind fuel is remarkably consistent. There are certain seasons of the year in any climate when the total amount of wind that will blow will fall within a narrow band of averages. In California, in the summer months wind turbines may register operating efficiencies approaching 90 percent! Still, day to day, hour to hour, the wind might be blowing or it might not. To deal with this intermittency, Heronemus envisioned bypassing the utility monopoly grids altogether. His ultimate solution to the energy crisis was to convert the kinetic energy of the wind into liquid hydrogen, which could then be pumped in pipelines to where it was needed for industrial, residential, or transportation uses.

A Heronemus mention of the great potential for wind power in Alaska prompted a state senator there to insert his 1972 technical paper into the *Congressional Record.* He become an instant media sensation and began giving utility and oil company executives bad cases of heartburn. Heronemus wiped the dust off a paper authored by Palmer Putnam that included the wind energy data for eight significant wind farms in the Green and White mountains of Vermont and New Hampshire. Heronemus expanded upon this initial work. He was soon confronted by one of the leaders of the Audubon Society after a public presentation. "If you put windmills in our national forests, we will put you out of our way," he claims to have been warned. Such public lands were no place for power-generating technologies, the nation's top bird watcher complained. A bird watcher himself, Heronemus began to factor the concerns of the Audubon Society into his enthusiastic plans for large-scale wind farms.

Shortly thereafter, Heronemus was offered a full professorship in ocean engineering at the University of Massachusetts at Amherst, despite lacking the traditional credentials and training. "In truth, I think they hired me because they thought I would bring in U.S. Navy funds," he chuckled. Unfortunately for him, MIT and the Woods Hole Oceanographic Institute soaked up all of the available research funds for his specialty, submersibles, so Heronemus pondered what else he could do. "As long as I could feed my graduate students and provide them with beer, I would be OK. And that required some research funding." he said. He observed that a similar desire to attract graduate students prompted Hutter to develop a wind energy program at the University at Stuttgart after the war.

Partly because of the concerns of his fellow bird watchers at the Audubon Society and partly because of his deep knowledge of the ocean's thermodynamics, Heronemus became enamored of the idea of building huge wind ships that could moor at sea, where the wind resource was far better than on land and where birds of any type were few and far between. According to Heronemus, the two best wind ship sites were the Aleutian Islands, off the coast of Alaska, and George's Bank, some 200 miles off the coast of nearby Cape Cod. He decided to start engineering programs at the University of Massachusetts in wind power and another one of his favorite fantasies, ocean thermal energy, a means of extracting electricity from ocean currents.

◆

Heronemus is still alive and is still pursuing funding for his idea. He was closing in on 80 years old when I first met him. On a good day, he could pass for somebody in his '50s. "Heronemus," barked a gruff voice on the end of the line. When he gave me directions to his house, he referred to landmarks on the "starboard" and "port" sides of the road. The curt chat ended with his "Righto!" He wore a set of large-lens glasses and was bundled up in a fat sweater that tied in the front when he answered the door of his farmhouse, with a red barn in the back, just on the outskirts of town. A curmudgeon, I'd been told, and reportedly a demanding teacher in the classroom, Heronemus seemed relaxed as we sat in the kitchen, sipping coffee. Large dark and wild eyebrows contrasted with his silvery short hair. "I'm going to ramble," he warned. "I'm supposed to. I was a college professor." He laughed freely, admitting his wife thought he was nuts.

Heronemus still lives just a few miles down the road from where he started the nation's first graduate class in wind power engineering at the University of Massachusetts at Amherst in the early '70s. The 50 or more students who graduated from his programs quite literally became the modern wind industry as they started up companies, worked in federal laboratories, or filled other positions in the private and public sectors.

The only work of his that came close to becoming a real wind farm was done for Molokai Energy, a new company he joined to develop wind projects at Ilio Point on the northwest corner of Molokai, one of the smaller, more sparsely populated Hawaiian Islands, at the start of the wind farming boom. "There is a remarkable, magnificent wind resource there," Heronemus enthused. His big-picture scheme was to generate the electricity and "send the power by cable to the other islands with all of the people." The whole project fell through when some Arabian interests purchased the Molokai Ranch Co., whose investments in new pineapple plantations failed and whose cattle herd was wiped out by disease. Though Molokai Energy owned the wind rights on the property, the plans for the wind project evaporated.

Heronemus claims that his wind ship idea does not require any radical new technologies. "It is a combination of ocean engineering, naval architecture, and mechanical, structural, and chemical engineering," he said. While the "correct technology has been conceptualized," he quickly acknowledged his wind ship ideas "still need to be verified by proper development and demonstration." Heronemus does not endorse any particular wind turbine design. His favorite, however, was actually conceived before World War II, in the 1920s, by Hermann Honeff, another German. It was far from light. The design, which was never constructed, featured clusters of wind wheels in the same plane adjoined to a 300-foot tower. Heronemus likes to think big. He envisions a total of 36 wind wheels, each measuring 62 feet in diameter, spaced appropriately on the legs of the tower, for ocean- or land-based wind power systems. Honeff's design was officially rated at 50 MW. At that rating, it would take 40 of the gigantic Smith-Putnam wind turbines to generate the same amount of electricity! Heronemus, however, claimed Honeff fudged the numbers, and the actual power output would have been significantly less. The key appeal of such large towers is simple, he stated: "The higher you go, the better the wind resource."

◆

Heronemus's bold ideas about wind power caught the attention of Lou Divone, an MIT graduate working at the National Science Foundation (NSF) who shared an interest with Heronemus in wind power. When a proposal from one of Heronemus's students crossed his desk, Divone read it with keen interest.

Fresh out of Viet Nam, Forest Stoddard, who goes by the name "Woody," was looking for a way to apply his helicopter engineering knowledge to design and build something that would make the world a better place. Born and raised in the Amherst vicinity, Stoddard was one of the first to enroll in Heronemus's class in 1971. "[Heronemus] is a bit like Bucky Fuller," Stoddard remarked. "Only instead of being radical, single-minded and political, he was a practical engineer. All of Bill's stuff is based on things that have been done before. It's not pie-in-the-sky. He helped me realize that I could help better the world with my engineering skills, that I could make a lasting contribution that would far surpass anything I did in Viet Nam."

I interviewed Stoddard on a rainy night in his office, a tiny room filled with computers, located in what used to be a church. Stoddard still plays surf guitar in a band called "Saturn and the Outer Planets," but he doesn't really look like a rock star. His glasses are quite large on his pale face, his thin red hair clings to his skull. He consumes at least three Cokes, a couple of cups of coffee, and perhaps a half pack of cigarettes during our three hour conversation.

Heronemus encouraged Stoddard to send out grant requests to underwrite the purchase of hardware for a few fantasy engineering projects they both had on the drawing boards. Stoddard sent the first of his proposals to NASF, asking for funding to design and build a wind turbine to supply power for a house on the University of Massachusetts campus. About a year later, the same time that Heronemus was making national headlines with his calls for major investments in wind power, the funding from NASF actually came through.

Divone wanted Stoddard to solve some of the technical issues associated with designing what was then considered a medium-sized wind turbine (but a small one by today's standards) whose power could be fed into the utility grid. "At the time, all that was out there were machines being imported from Australia which could charge DC batteries or, with some retrofitting, could be connected to the AC grid," Stoddard explained. The largest of these turbines were rated at 25 kW. That's just enough electricity to power a typical house, noted Stoddard.

Then came the oil shock. The price of petroleum jumped 70 percent in October of 1973. Divone quickly helped convene a wind power workshop that featured Heronemus, Putnam, Marcellus Jacobs (who still pushed wind power well into his 80s), and Hutter. With the energy crisis fresh in the minds of policy makers in our nation's capital, a sudden surge of support rose to finance alternative energy sources, and Divone was the key figure in doling it all out as he moved from NASF to become the point man on wind energy for the federal Energy Research and Development Agency, later reconstituted as the Department of Energy (DOE). Divone authorized funds for the refurbishing and retesting of the Gedser mill from Denmark. Data collected from this mill served as a baseline of aerodynamic and structural dynamic performance information for both the U.S. and Danish wind industries. Divone then turned to NASA's Lewis Research Center for technical assistance, since U.S. engineers there were measuring winds in Puerto Rico and there was some local interest in pursuing a wind project. In 1974, the federal government launched its multimillion-dollar wind energy program, dividing its dollars among national labs, with each lab focusing on particular aspects of wind power: Rocky Flats was dedicated to small turbine development; Pacific Northwest Laboratories, which had the best meteorology expertise, mapped the nation's wind resource; Sandia Labs focused on development of vertical-axis turbines; and so on.

Since electric utilities—identified in studies conducted by companies such as General Electric (GE) and Lockheed as the prime market for large wind turbines—still viewed wind turbines as a renegade technology, they were less than enthusiastic about sinking any of their own funds into its development. However, GE and Lockheed, with half-million dollar DOE research contracts from Divone, saw incredible business opportunities ahead. Each independently confirmed that Heronemus was right. America possessed large amounts of wind that could be farmed to generate electricity. They estimated that as much as 18 percent of the nation's electricity could easily come from the wind if turbines at least as large as the Smith-Putnam were grouped together in "farms" and, like traditional central-station power plants, fed their electrical output into the utility grids. Though this vision echoed Heronemus's, there was one main deviation: Heronemus wanted these wind farms to bypass the utility grid and produce hydrogen. GE and Lockheed instead suggested that utilities become the primary consumer of wind-generated electricity. GE and Lockheed also dismissed Heronemus's offshore recommendations. Several terrestrial sites had enough wind to generate electricity cost effectively, they maintained. They agreed, nonetheless, with Heronemus that the Great Plains was the nation's prime wind resource area.

Though the tiny wind farming community, primarily Midwesterners, welcomed studies confirming there was plenty of wind blowing in the United States to generate huge amounts of electricity, they were outraged by the proposed government focus on gigantic wind turbines feeding gobs of electricity into a grid that had nearly killed their parents' or their own wind farming businesses. After all, most wind farmers viewed wind as an alternative to business as usual, a way to control one's own power and avoid plugging into the grid. In their minds, the wind turbine was a technology that needed to be developed at a scale appropriate to tasks at hand with some recognition of the local environment. The ideal future scenario consisted of thousands of small wind turbines installed at individual homes, farms, and businesses by small local firms.

Hans Meyer, a disciple of Bucky Fuller and the head of a small wind power company, said this at a DOE workshop in 1975:

> There are many serious questions about the whole philosophy of centralization of our energy system. Because we have been moving in that direction for 75 years does not necessarily imply that we should continue. . . . If there is an energy crisis, it is one of consumption, not of production. We need to learn better how to live on our spaceship earth. . . .

The DOE studies suggested that the biggest bang for the buck would be in gargantuan turbines three times the size of the Smith-Putnam machine. Two-bladed rotors facing downwind—the Smith-Putnam configuration—made the most economic sense. (For comparison, the Jacobs and Danish machines were three-bladed and faced into the wind.) Relying on these studies by companies that had never designed or manufactured a single wind turbine, Divone began to marshal political support for a Federal Wind Energy Program (FWEP) that would represent the world's most aggressive attempt to parent an infant industry.

The DOE/NASA wind turbine development program was launched in 1975. Divone started out with a staff of only five people: one meteorologist, a few engineers, and a secretary. He pushed for the largest wind turbine rotors possible, at one point calling for a 400-foot fiberglass rotor. The largest rotor in existence in the United States was only 49 feet and was used in a helicopter. The first DOE/NASA turbine, the MOD-0, featured a 200-foot rotor, which exceeded the wingspan of a Boeing 747. Later, a 300-foot rotor that fell just short of matching the wingspan of the largest plane ever built—Howard Hughes's 320-foot "Spruce Goose"—was used on a DOE machine.

The results from Divone's program were hardly encouraging. A case in point is the 1976 awarding of a $7 million contract with General Electric and United Technology to develop a 1.5 MW wind turbine based largely on the MOD-0 design that was the apple of Divone's eye. The resultant design was not a success. It generated only 30 hours of electricity over its 10-year life. Out of a total of $24.5 million disbursed by the federal government in 1977, more than $17.5 million went to such names as Lockheed, Westinghouse, NASA, Boeing, Grumman, and GE to build bigger and bigger wind turbines.

An animated man seemingly always on the move, Divone sat, at least for the moment, in his crowded Washington, D.C., office filled with photos of the wind turbines that he helped build over the years. He grips my hand with such a firm handshake it hurts. Divone, whose most memorable features are his wide frame and large nose, is full of stories about each and every one of his beloved DOE wind turbines.

When I visited him shortly before his retirement from DOE, Divone acknowledged his disappointment with the quality of work on large turbines. He chastised the aerospace manufacturers for the very visible failures of machines trumpeted as being commercially ready:

> We've had a lot of complaints sometimes about the "aerospace" influence, but I haven't always seen aerospace quality. There are equivalent horror stories about intermediate and large machines and cold solder joints, repetitive failure of hydraulic systems and bearings and controls and on and on. I'm not sure yet how we're going to handle this problem or what's going to be done. . . . I'm getting just a little tired and a bit angry, but I feel that wind energy as a whole has come too far to take a risk of failure now.

Several big MOD series turbines were erected in Hawaii, Washington State, and elsewhere, but the envisioned third-generation technologies never came to be because of the election of Ronald Reagan and the subsequent shift in values and priorities at DOE after 1980. The only federally financed wind turbine installed in California was near the Altamont Pass, in the same Diablo Range, but on an eastern slope. Installed by Pacific Gas & Electric, the MOD-2 was a two-bladed monster designed and built by Boeing. Jacobs told DOE early on, in no uncertain terms, that the machine would never work. And he was right.

The wind regime at the site was incredibly turbulent, and flow patterns severely challenged a rotor as large as the one on the Boeing machine. At an altitude of 200 feet, the winds might be out of the southwest at 25 miles per hour. At 310 feet—the height of the tip of the blade as it reached up toward the sky—the winds might only be 5 mph out of the northeast. This large, $10 million lemon was the last of the DOE-financed machines and was one of its more spectacular failures, producing only 40 percent of the power expected from normal consistent operation. This was perhaps the clearest example of how DOE's ivory tower and theoretical approach to large wind turbine development was divorced from the demands of real-world wind regimes. Rather than dismantling the beast and selling the turbine parts as scrap, PG&E used dynamite to blow it up.

Trying to make decisions about where to spend taxpayer dollars was difficult, Divone acknowledged, because "all of the economics were flaky." The numbers and projections being bandied about were pulled out of thin air. "We all suspected that economics was a function of size," Divone remarked. He showed me a few diagrams, hastily sketched out with pencil on paper, that demonstrated the rationale behind the federal government's focus on building bigger, and allegedly better, wind turbines. He claimed that the margins on a small wind turbine are so thin that a single maintenance cost can wipe out a year's worth of energy. The larger you make the machine, the lower the cost of producing the energy, and the better chance at long-term profitability, or so said his line on a piece of paper. But, he pointed out, as you build bigger machines, the capital costs of components starts going up. The simple devices for yaw control in a lightweight Jacobs machine cannot withstand the turbulence when scaled up 10 to 20 times. The answer to this dilemma, according to Divone's zig-zagging diagram, is to "build bigger to cover the cost of those widgets, and the cost of energy then starts dropping again." But then "the square cube law catches up again." Divone was referring to an axiom mentioned in Chapter 1: A doubling of wind speed results in an eightfold increase in potential energy production.

Unfortunately, stress on the hardware makes an equally huge leap, driving potential profitability back down. In essence, Divone was plotting the shape of a "W" in his economic projections relating turbine size to operating profit. Neither Divone nor anyone else at DOE knew what real-world numbers fit at all the ends of his zigs and zags. Nonetheless, this diagramming exercise had convinced them that they needed to build wind turbines the size of the Statue of Liberty. With an influential constituency at government labs and the large contingent of U.S. aerospace and military weapons firms looking for new pork barrel projects, research on behemoth machines was quite appealing and soaked up the vast bulk of Uncle Sam's investment in wind power.

Although the shrewdness of most of Divone's DOE investments may have left something to be desired, the small wind turbine Stoddard designed with Divone's money for Heronemus's wind power class showed some promise. Working closely with Ted Van Duzen, another Heronemus graduate student who had experience designing composite materials in the Navy, Stoddard took five years to build it. The turbine was rated at only 25 kW, but it appeared to be reliable in the relatively light winds at Amherst. The project became a very visible lure to attract graduate students to the only wind engineering class in the Unites States. The renown of Heronemus and his optimistic views about wind power spread quickly among a tiny, but growing, group of engineering students fascinated by the challenge of converting wind into electricity. Soon his classes at the University of Massachusetts attracted students from as far away as Europe.

Chapter 7

Small Is Beautiful

> Only people serving an apprenticeship to nature can be trusted with machines.
>
> —Herbert Read

William Heronemus had a daughter attending Wheaton College in the mid-'70s. Two doors down from her dormitory room was the daughter of Russell Wolfe, a soft-spoken engineer harboring a deep idealism. After hearing Heronemus's daughter go on about her father's visions of wind power, Wolfe's daughter chided him to do something in his life as worthwhile as developing renewable energy sources. Perhaps as a reflection of the times, Wolfe responded to his daughter's challenge not with disdain but with curiosity. She prompted him to take a close look at wind power. Perhaps this was a business opportunity that could dovetail with his own gung-ho views about the ability of technology to deliver great value to his pocketbook and the environment.

Two weeks later, Wolfe showed up at Heronemus's office and spent the whole day talking with him about wind power—how it worked, how much wind fuel was really available throughout the country, what was the current status of the technology. Wolfe then approached his friend Stanley Charren about starting a new business venture. Charren had helped him out in the past and they had become pals. If there was somebody out there who had the money to invest in new technologies, Charren could make it happen, Wolfe reasoned.

"The first time I got a phone call from Russell, he told me about a technology he invented that could measure high vacuum pressure," Charren recounted. Charren was a principal of a Small Business Incubation Corp., an SBIC. SBICs were quite popular in the '50s. Those interested in pooling their fiscal resources in order to help launch new companies could create a SBIC and gain a few tax advantages. The SBIC Charren helped establish was composed of "technical weenies" looking for new technologies that could fill an immediate niche and turn a quick, sweet profit. "New England has always been a hotbed of technical innovation—particularly Massachusetts," said Charren. "Back then Route 128, and the greater Boston area, had a reputation as being an area like Silicon Valley, a place where something could happen." Charren helped Wolfe set up a company to market his pressure-measuring device, and it was a financial success.

Charren is no stranger to success. He made a nice bundle in a business that marketed simulated leather products. Among his other notable business ventures was a new kind of carpet company that developed removable floor tiles that could be recycled instead of being thrown away. This firm evolved into the world's largest floor covering company; now called Interface, Inc., it is considered one of America's "greenest" businesses by virtue of its waste-reduction and recycling practices and its recent purchases of renewable energy including solar and wind power. Schooled at Brown and Harvard universities, Charren looks and sounds like a classic Bostonian, one who spends much of his summers at Martha's Vineyard and Cape Cod.

"Russell knew I had some training in aerodynamics as a graduate student. I had done some work on airfoil configurations and propellers—so he knew I would understand the concept of wind power," Charren said. "I was skeptical. But I knew Wolfe was a trained scientist. And he had the insights. He showed me some really good numbers about the potential for wind power." Wolfe put Charren directly in touch with Heronemus, the source of the numbers showing that America was a gold mine when it came to wind fuel.

Charren made a small contribution to the U-Mass wind energy program and then, sensing that large profits were virtually blowing in the wind, offered to employ Heronemus in a new wind power company. Though intrigued, Heronemus turned the employment offer down. But his two prize students, Stoddard and Van Duzen, accepted, and in 1979 a new company was formed, with Wolfe and Charren as the other two members.

The company, U.S. Windpower (USW), had a strategy that was unique for its time. Whereas DOE focused all of its efforts on building complex, colossal machines, most of the private industry had focused on tiny turbines in the 1 to 5 kW range and designed for individual residences or farms often not hooked up to the grid. The strategy that Charren hit upon was to focus on an intermediate machine, one that sat between these two extremes, and to connect these turbines to the grid and sell the power to utilities. Stoddard's small 25 kW wind turbine design, having started out as an experiment for class, suddenly became the technological basis for a new company hoping to show that bigger was not necessarily better.

From the start, Charren's strategy was to rely on a "windplant," the company's trademark term for a collection of numerous medium-sized machines controlled by a computer communications system in such a way that each individual turbine operates as if it were part of a single power plant. The idea was to maximize production. A key reason this deployment scheme made sense was that each turbine was responsible only for a small portion of the total output. When individual turbines were shut down for maintenance and repair, which Charren recognized would occur frequently with prototype machines, the windplant as a whole would incur only small revenue losses.

The first public offering in 1978, for $1 million to finance development of Stoddard's wind turbine, failed miserably. The concept of wind power had been around for a while, but the nature of the wind—its random intermittency—still gave it a stigma among many potential investors. However, soon thereafter new federal laws offered incentives for investors and required electric utilities to purchase power from a new class of market players that would come to be known as the independent power industry. The Three Mile Island incident had cast a pall over the nuclear industry, and alternatives such as wind power were now being given a fresh look.

Charren started bringing some new blood into the company, mostly cronies who were recently retired corporate bosses and were looking for an exciting opportunity to sink some capital into what was being billed as cutting edge stuff. Two critical additions to the fast-growing roster of the company were Norman Moore, a former president of several divisions of Litton Industries, and Herbert Weiss, who made a name for himself at MIT's Lincoln Laboratory for inventing a distributed early warning radar system, called the DEW Line, inspired by the Cold War's preoccupation with the Soviet threat.

Moore was installed as president and wrote a "red book" that was to be used to raise more than $5 million for the small, struggling company. The book pointed out that USW was taking a vastly different approach from that of DOE, which "concentrated upon the development of large megawatt-sized turbines assembled in aerospace factories typically using methods and components developed in the capital goods rather than consumer goods industry." USW had instead "designed a windmill which is about the size and weight of a small car, capable of utilizing the techniques of mass production of the consumer goods industry, and achieving megawatt output by combining many medium sized turbines rather than depending on the output of one very large machine." Moore admits the book contained its fair share of hyperbole and tall talk about the potential profits, potential partners, and the readiness of its wind turbine technology. But it worked. Money started coming in, primarily from wealthy New Englanders who were friends or acquaintances of Charren or Wolfe or Moore.

Weiss was put in charge of engineering, even though he had absolutely no experience in wind technology. Stoddard saw Charren's buddy Weiss as a johnny-come-lately who bought his way into leadership of the company by virtue of his own financial resources. Stoddard, whose acerbic tongue often left no room for interpretation, soon began to feel that USW was no longer a tight-knit group of idealists whose destiny rode with what he and Van Duzen could come up with in turbine design. He had been motivated by the picture of our energy future articulated by his mentor Heronemus; now, he started realizing that business in the business world still meant business. Stoddard was quite possessive of this little start-up company and resented the new influx of partners from a culture vastly different from his own. They had not known Heronemus and had never been inspired by his boundless idealism and commitment to sound engineering principles. Stoddard began to sense that the organization's culture was shifting.

The first meeting of the new board of directors Charren had put together and the new three-piece-suit investors Moore had attracted to USW was held in 1979. At the start of the meeting, Moore unexpectedly turned to a new hire, Louis Manfredi (another of Heronemus's graduate students), and instructed him to start up the latest iteration of Stoddard's prototype wind turbine. Manfredi had just had a nightmare in which a turbine blade broke off and sailed straight up into the sky and then came smashing down and collapsed the other two blades. Since the turbine had never operated for more than a minute or so, Manfredi protested to Moore that he didn't think starting it up was a good idea. Moore insisted. "It's their money. Let's see where we are," he said. What ensued was Manfredi's nightmare: The first 15-foot-long fiberglass blade snapped, shot into the air, and came crashing down on the other two blades, the whole mess falling to the ground with a loud, reverberating thud. Luckily for USW, the company's first major outside investors did not seem overly concerned.

But Stoddard was furious. He had told upper management that the machines were not ready for prime time. "I can't in good conscience say these wind turbines are safe," he warned. Yet he felt that Weiss kept increasing the pressure to design a wind turbine that would perform and fulfill Moore's grandiose promises to the investors. Designing a wind turbine to consistently respond to the wind's whimsical disposition had turned out to be a much bigger puzzle than ever expected. It took Stoddard four shots to come up with the current prototype. The first version worked only five minutes before falling apart. Subsequent prototypes faced similar fates, disintegrating before the eyes of Stoddard and an increasingly impatient Charren.

Among the problems Stoddard identified were the following: The cast aluminum part that connected the fiberglass blades to the rotor hub by epoxy failed during extended operation; the six large bolts that attach the rotor hub to the drive shaft failed, as did the pitch control system, which relied on tie rods similar to those used in cars; the interior temperature of the nacelles was higher than expected, reducing generator efficiency; the microprocessor software often turned machines off when they could have been operating, thereby reducing power production. For each of these problems, Charren and Moore assured potential investors that a solution had been identified and was being implemented.

In Stoddard's eyes, however, USW was being transformed from a small idealistic group on the frontiers of knowledge into just another big profit-seeking corporation that would sell snowballs in hell if it could. At the next board meeting a few weeks later, the latest Stoddard design was turned on again. Again, the blades started flying all over the place. This time the failure was more ominous. These models were already being installed at the company's first wind farm at Crotched Mountain, New Hampshire.

As a public relations ploy, Wolfe had grabbed headlines by claiming that USW was building the Crotched Mountain project as a way to create jobs for unemployed mill workers in nearby Lowell, Massachusetts. He touted it as the source of electricity for the Crotched Mountain Rehabilitation Center, an institute for disabled children. The site was perfect. The average wind speed was an alleged 20 mph; roads to the site already existed; and the land was within reach of utility transmission lines. The rehabilitation center was also happy to trade use of its land for the wind-generated electricity. The fact that there was no practical way to trace any of the small amount of electron flow these machines actually produced to the alleged recipient of this clean electricity did not deter Wolfe from bartering.

"Wind farming will become a secondary source of income for farmers and landowners," he predicted after the project was up and running. "A sort of second cash crop. But building a hundred turbines is at the low end of feasibility for an efficient wind farm. We're talking of thousands at one site. Having twenty turbines on Crotched Mountain is like trying to raise wheat on three acres." There was no need "to wait for any major technological breakthroughs. The technology of wind power is so well established that six guys in a garage can make a living at it. People will buy wind power from a wind farm not because it is a wind farm, but because it is cheaper power. At some point, the price of wind power is going to determine the price of oil."

Wolfe's boundless enthusiasm for wind power technology notwithstanding, the blades soon starting snapping off at the Crotched Mountain site. Stoddard grew despondent. Weiss had given him an ultimatum: Get these machines working by spring of 1980. According to Stoddard, Weiss then left on a six-week cruise.

The tension between satisfying investors and developing a wind turbine that would actually stay in one piece to consistently generate electricity weighed heavily on Stoddard. In a letter to USW management, he complained that the most recent prototypes that Weiss had designed were steps backward, not forward, "and are clearly indicative of a management with only cost and rapid movement as goals." He described the new structural requirements, dictated by Weiss to further reduce costs, as "a joke, a toy" at the expense of design integrity, and another Weiss-approved design "as a wild guess, by committee, that was foisted upon me through devious back door means." He summed up his frustration, complaining that "by focusing on cost goals to the exclusion of everything else, the machinery and design are rapidly becoming unsafe and inferior."

Stoddard also lamented that upper management had "consistently failed to approach, motivate or understand its workers. This is a puzzling occurrence in people whose careers have been so successful. In this company, there is no silent middle management to take up the slack, mold the unity of purpose, and buffer the employees. This management has had many opportunities to do this and has consistently refused to listen. This management appears suicidal; I happen to believe that the goals of the original company were worth sacrificing for, were worth working for and building for. I intend to do all I can to save this before it is too late."

But the solutions were not to come from Stoddard or Van Duzen, who were suddenly booted out of the company as Charren grew tired of all the carping between upper management and the two original design engineers. The final straw came when Stoddard demanded that production of Weiss's design be halted. Perhaps the students of Heronemus had outlived their usefulness. The company now had a machine. It had to concentrate on marketing, and that task was becoming increasingly difficult when people within the company kept talking about all the problems.

Ironically, the Crotched Mountain installation of twenty of USW's 56-50s garnered significant press. The alluring story of farming the wind captured the imagination of a general public seeking appealing solutions to the nation's energy woes. Less press was devoted to the fact that within a year, all of the wind turbines had been destroyed and the entire site bulldozed over. The skeletons of the company's "commercial" wind turbines were buried below mounds of dirt.

Part II

Chapter 8

Outlaws of the Altamont

Madame, bear in mind that princes govern all things—save the wind.
—Victor Hugo

The aluminum foil sky radiates intensity in the Altamont Pass this August afternoon. The cool winds off the Pacific sweep over smooth round bumps, rushing to fill the void left by the rising heat of California's Central Valley. Undulating fans of soil unfold in each and every direction as if liquid earth had been squirted out of a large tube. This landscape of ho-hum hills, some 50 miles due east of San Francisco, turns gold every summer. During this time of the year, a steady stream of winds from the west turns the white propellers planted on narrow, bald ridgelines, generating electricity for the masses while lining the pockets of a few lucky profiteers.

Randy Tinkerman points in the direction of the pinwheels that populate this northern stretch of the Altamont Pass, which lies on the southern fingers of the modest Mount Diablo coastal range. I follow him and three others up a trail on a clandestine tour of wind farms that I will never forget. Tinkerman knew the combinations to the locked gates of local ranchers and therefore had access to private properties in the Altamont that are off-limits to tourists and most wind farmers alike. Hung over, but pumped up on coffee as strong as midnight on a new moon, I couldn't help but perk up as I came into the midst of arrays of wind turbines, spinning furiously all around, generating whooshing and whirling sounds, a kaleidoscope of moving arms everywhere. As we hike, perspectives on the machines constantly change, rendering new patterns of fluctuating blades—some skinny, some fat—that keep capturing my attention.

The kinetic energy of the wind rushing in from the Pacific rotates white propellers as far as the eye can see. Most of the wind turbines scattered along the ridgelines here sport what are essentially helicopter blades rotated 90 degrees from the horizontal plane to the vertical. I feel mesmerized by this vivid picture of a technology planted in a landscape where it commands attention and so clearly makes the persistent wind so palpable.

◆

The first residents of the Altamont Pass were the Costonoans, whom early settlers thought were an entirely different race from the Indians they had encountered on the other side of the Sierra Nevada to the east. Their timidity and fondness for nudity perplexed early Anglo visitors to the region. "I have often seen hundreds of them grazing together in a meadow like so many cattle," writes an author of a book published in 1883. (It is unclear whether the author witnessed these Indians crawling on all fours or not.) But many of the stereotypes of Californians that persist in the media today—the laid-back, mellow relaxed dude with an aversion to too many clothes—may be perpetuating an image first established by indigenous peoples. What is now Highway 580 used to be one of the prime trading routes of Costonoans, Yokuts, Miwok, and others. Tribes from the east met tribes from the west to trade obsidian, ochre, acorns, and abalone. Unlike most of this 80-square-mile wind farm area, this particular spot where Randy Tinkerman and I now stand includes a number of other special caves that offer protection.

"Many of the caves here are covered with Indian art," says Tinkerman. He hired an archaeologist to evaluate the intricate, yet raw drawings. "Are they 200 or 2,000 years old?" Tinkerman wondered. "If they are 2,000 years old, they have a remarkable presence. If they are 200 years old, they are fading fast." The archaeologist gave inconclusive answers. But Tinkerman has a theory. As he gazes off into the distance, the turbines chopping up the red horizon, he spins the following scenario linking the Altamont to one of the darker portions of California's past—the persecution of Native Americans by overzealous Catholic missionaries. "There was an uprising. The Indians made an escape from the mission. Where would they come? Where they least likely would be—the Altamont Pass! The escapees from the mission may actually have lived here. Instead of just meeting here to trade, maybe the coastal and foothill Indians would come here and stay sometimes."

Tinkerman noted that Native Americans also took advantage of the Byron Hot Springs, which are located nearby. They traveled as far as 200 miles to be healed by the mud baths and hot, bubbling waters. Tinkerman's face suddenly tightens as a long pause passes. A quick look down. The vertical turbine structures, and the long reach of the horizon, produced a powerful, and hypnotizing, image. "Big cave rooms. It's magical to hang out there. It's a very strange environment. Nothing like these other hills."

A large raptor suddenly sweeps into view. Viewed through binoculars, it appears to be an immature golden eagle, its age discerned from the white marks under its wing and tail. It is soaring just a few feet above the ground, swiftly moving along the contours of an unbroken ridgeline. It dips a wing and pops over into the next vale, snaring a shrieking ground squirrel. Eagles, hawks, falcons, kestrels, and vultures all frequent the Altamont Pass and rely on the subtle nuances of prevailing wind patterns to hunt prey easily spotted on the nude hills. The Altamont Pass being located on a major migration route, its resident birds are joined by thousands of passersby in the spring and fall as raptors soar and swoop, riding the zephyrs, foehns, and other wind forms that frequent this seeming paradise for birds on top of the food chain.

The Sacramento Valley ceremonial system dominated the myths of the Native American tribes that populated this region. Most of these tribes worshipped Mount Diablo, a peak that rises close to 4,000 feet above its sea-level, pancake-flat surroundings and anchors the coastal mountain range that includes the Altamont. Tribes that have lived here were not nomadic; generation after generation lived in the very same spot, until driven from their homes by the whites. Among the animal deities that reigned over the Altamont were Coyote-Man, Condor, Turkey Buzzard, and Golden Eagle.

Early settlers never bothered to figure out the myths and legends of local tribes. To them, these native peoples were "lazy and filthy." One Alameda County historical document put it this way: "Nature had provided for them with a lavish hand, and all they had to do was to reach forth with their hands, pluck and eat. No vain ambitions lured them on in the great race of life; no baubles of riches enticed them into hardships of labor, either mental or physical. They lived to die." This official county record went on to note that within a half century virtually all members of local tribes had disappeared: "Cholera took them by the thousands in 1833. . . . [I]t was said they died so fast that the living were unable to care for the dead." And while disease had a devastating impact, civilization took an equally heavy toll. "Then came war [the Civil War] with its kindred calamities as another decimator of their ranks. Soon after the whites came among them, prostitution became general; the women no longer bore children, and thus the tribe gradually, but surely, died out, and no little ones grew to take the place of the deceased elders. Truly would it appear to have been a matter of destiny, for it was impossible that the two races could exist in contact."

The only cure local tribes could offer for disease was "sweats"—rituals of cleansing. A fire was built in an enclosed chamber and the afflicted literally sweated out impurities that had become stored in the body. Sweats were practiced frequently near the Altamont. The Anglo settlers who witnessed these ceremonies of spirit found them quite scary:

> . . . round about the roaring fire the Indians go capering, jumping and screaming, with perspiration streaming from every pore.

> The spectators look on until the air grows thick and heavy, and a sense of oppressing suffocation overcomes them, when they make a simultaneous rush at the door for self-protection. Judge their astonishment, terror, and dismay to find it fastened securely—bolted and barred on the outside. They rush frantically around the walls in the hope to discover some weak point through which they find egress, but the house seems to have been constructed purposely to frustrate such attempts. More furious than caged lions, they rush boldly against the sides but the stout poles resist every onset. . . . the uproar finally ceases and the Indians vanish through an aperture opened for that purpose. The half-dead victims to their own curiosity dash through it like an arrow, and in a moment more are drawing whole bucketfulls of the cold, frosty air, every inhalation of which cuts the lungs like a knife, and thrills the system like an electric shock. . . .

The Altamont wind farms link the present and past. The families that still reside here, ranchers for the most part, are often descendants of the first white settlers to eke out a living on these godforsaken hills. Chances are these ancestors helped decimate the Native American populations that worshipped the golden eagles of the Altamont Pass.

One of the first recorded observations of this region was that of the Juan Bautista Anza Party in 1776, the year the 13 colonies declared their independence from the English crown. California, that wild, distant land on the opposite side of the continent, was still firmly in the hands of Mexico. By 1839, the Altamont Pass was commonly referred to as "Sierra de las Buenos Aires"—Mountain Range of Good Winds.

The Mexicans who inhabited California before it joined the union in 1850 had called the Altamont "Sierra del Chasco"—Mountain Range of Jokes. I have been unable to find any explanation of this provocative title. I suspect it may have something to do with either the constant wind, perhaps suggesting voices telling jokes, or that the range itself was a joke when compared to a mountain.

Brushy Peak is the highest point in the Altamont Pass. Its few snarled trees look like hunched-over creatures with scraggly appendages. Beneath their sharp root claws lies a secret cave. Local legend has it that Joaquin Murrieta hid much of his gold and other treasure here, in a land frequented only by crooked, howling winds. Murrieta was the classic Mexican bandito who roamed California, robbing forty-niners of their gold throughout the mid-1800s. Though the entrance is quite small, Murrieta's alleged secret room in the Altamont Pass is the size of a parlor and includes two seats. From here, the view of the spinning wind turbines becomes mesmerizing.

Murrieta's gang rounded up horses throughout the Sacramento Delta region and corralled them on the eastern slopes of the Mount Diablo range. The Altamont Pass was the northern boundary of the area where his banditos robbed and looted on trails that stretched south to his native Mexico. Starting in the gold country in the Sierra Nevada, where rival Anglo gold miners first abused him because of his ancestry, Murrieta and his minions soon claimed victims in Bakersfield, Santa Barbara, and other spots up and down the state. In 1853, he was captured and allegedly beheaded, to the relief of many. However, reports of Murrieta still kept popping up, leading some to say his ghost lived on. Others claim he was just a hoax, a myth, yet another invention of that unique California imaginative frame of mind.

The namesake town of this region—Altamont (literally, high peak)—was founded in 1868, when the Southern Pacific Railroad linked the cities of Oakland and San Jose with Sacramento by laying its track on what used to be the Old Spanish Trail, a road that was frequented by Murrieta and his gang. The trail dates to 1772, when early missionaries searching for the mouth of the San Joaquin River blazed it. Altamont is now a ghost town. Whereas cities such as Livermore and Tracy sprouted up on either side of the Altamont Pass, the hills themselves are primarily grazing grounds for cattle. In the 1880s, real estate agents promised prospective landowners that lush fields of crops would fill the lovely hillsides. The unrelenting wind broke these promises.

Today, Hell's Angels roam these bare Altamont hills on their Harleys. I once saw over a hundred of them congregating at a rustic bar at the outskirts of the wind farms. From there you could see the spinning white turbine structures, both up close just beyond some cottonwood trees, and on ridgelines in the distance. The bar's summer patio featured tables that, in a previous life, were large spools for electrical cables that hooked the wind turbines into utility grids. The bar served only one beer: Pabst—in cans. It turns out a lot of the early wind turbines in the Altamont Pass were installed by Hell's Angels, a sign of how ubiquitous this tribe is here.

◆

The only event at the Altamont Pass to generate a blip on the general public's radar screen was a December 6, 1969, concert by the Rolling Stones which involved the Hell's Angels. Over 300,000 people showed up to see a concert that was billed as "the Woodstock of the West," creating a huge traffic jam in a windswept no-man's-land where silent slow-moving cattle are far more common than human beings.

The show was a holiday gift from the Stones to its growing legions of fans. The location kept changing, however, as the authorities' worries about swelling crowds foreclosed traditional free outdoor concert venues within San Francisco's city limits. Finally, rock promoter legend Bill Graham found a race track in the

Altamont Pass owned by Dick Carter, who thought the site was ideal for rock 'n' roll shows. It was close to San Francisco and Berkeley—and out in the boonies—all at the same time. Given the long search for a site, Graham rushed to put the show together in just three days. The resulting lack of planning led to tens of thousands of dollars in property damage to ranches in the region. But it got worse, much worse.

Perhaps the best insider account of how the vibes at the Altamont Speedway site spooked the Stones concert is from Owsley Stanley, the chemist who provided the early "Acid Tests" for San Francisco's psychedelic scene and is now quietly installing small wind turbines in Australia:

> It was like a moonscape of crushed auto bodies. Like just all crunched and crushed. As we drove along, we looked over to the left and saw this place that looked like a skull. It was the actual arena in which they held these demolition derbies. I thought, "Oh, my God. This place smells of death. Of the energy of people who come here to watch people crash these cars together, hoping they'll die." I thought, "This is the worst possible place to hold something like this." And I realized that if you took acid at the show, you were going to have a trip that you didn't really want.

Sensing disaster, the Grateful Dead, one of the acts expected to play, never even plugged in their equipment. Jerry Garcia commented, "It was in the air that it was not a good time to do something. It was too weird. And that place. God. It was like Hell." Shortly thereafter, Grateful Dead lyricist Robert Hunter penned a song about the concert entitled "New Speedway Boogie." Its chorus, sung by Garcia, is, "One way or the other, this darkness has got to give."

The Hell's Angels, the gang of leather-clad outlaws who roam the Altamont on Harley motorcycles like a nomadic tribe, were hired to be stage guards in exchange for $500 worth of free beer. The result was mayhem. The Angels poked at the crowd with pool cues, and when eighteen-year old Meredith Hunter, a ladies' man sporting a huge Afro, allegedly pointed a pistol at Stones lead singer Mick Jagger, they stabbed and then stomped him to death. Another man, freaking out on LSD, drowned in the California Aqueduct. Two other folks were run over as they were dozing in their sleeping bags. "If Jesus Christ had been there, he would have been crucified," commented Jagger after pop journalists tagged the show one of rock 'n' roll's darkest hours. "For me, it became the point my entire generation's hope turned to a reluctant resignation of how the world really is," said John Roberts, a T-shirt maker from Berkeley. "That was the start of the narcissistic 'Me' generation," he shrugged. "And the end of peace, love and all that, man."

Carter blames the media for making the Altamont concert a symbol of everything that ever went wrong after the Summer of Love. "Things were going so well, we talked about doing a three-day concert once a year. Those Stones were a bunch of snot-nosed little kids who didn't know what they were doing. They were just having a ball and playing with the drugs and stuff, and didn't stop to think that you have to do some real planning for something like that. Heck, you put together a half-million Catholic priests in a field in the middle of nowhere like that, and you'd have trouble, too." While three babies were born at Altamont, thirty people with injuries had to be airlifted to hospitals. Carter still can't believe how the day unfolded. "We put in 1,000 portable toilets, 2,000 garbage cans, rented 16 helicopters—but all day long they just kept coming, like ants over the hill. It was incredible."

Anglo settlers from nearby towns like Tracy held concerts at the turn of the century in these caves that Tinkerman cherishes, which lie across the freeway from the racetrack in the northern portion of this sprawling mega wind farm. Nevertheless, the only live music to infiltrate the bald hills of the Altamont since the Stones opened the show with "Sympathy for the Devil" as the chilly winds started to blow in from the Pacific was a string of groovy all-night "rave" parties in the mid-'90s. There is no music here today—except that generated by the wind.

Chapter 9

California Dreaming

Dreaming men are haunted men.
—Stephen Vincent Benet

"It was here on this very spot where my VW van broke down," chuckled Tyrone Cashman, a pale sliver of a man with silver hair. He has an air of wonder about the world around him as he glides across the room in socks, giving me a large mug of Chinese green tea. Cashman was telling me about how he ended up in his current home on the southwestern flank of Mount Tamalpais, in Marin County, worlds away from the blue bloods in Boston and the bad vibes of the Altamont Pass. If not for Cashman, few of the wind farms that still spin today would have sprouted from the ground in California. "When I arrived in California back in 1977, I, and a lot of others, believed a new world was possible. We had the confidence that we could do something really different—and do it quickly," he said, his eyes growing bigger as he reminisced. "Oil was going to run out and that would provide the price signal that would make alternative energy—such as wind power—competitive." He is not alone in still waiting for oil to rise to the $100-per-barrel price forecasts put forward by experts from outfits like the CIA.

California became a magnet for the wind farming industry because of public policies crafted by Cashman. Only in California, and only in a Jerry Brown administration, would the job of developing wind energy policy fall into the lap of a Catholic Buddhist philosopher who also just happened to be an avid student of history. After studying to be a Jesuit priest in Wisconsin and then in California (where he bumped into Jerry Brown), Cashman ended up at the New Alchemy Institute in Woods Hole, Massachusetts, in the mid-'70s. The Institute was a think tank, dominated by anarchists all dedicated to supporting small-scale technologies that walk lightly on the earth. He became enchanted with wind power, believing that it was a technology that offered solutions to the pollution so embedded in traditional forms of industrial growth.

Cashman ended up spending a lot of time at the institute working on a 25 kW wind turbine for New Alchemy clients in Ottawa, Canada. John Todd, who at the time was head of the Institute, was a sailor and, in the words of Cashman, "a native genius" who wanted to prove that wind power was an alternative to nuclear power. Utilizing his knowledge of how the wind works on sailboats, Todd put together a wind turbine. "Hanging around Todd for a whole summer tinkering with the wind turbine, I began to learn the language and philosophy of engineering. Since none of the parts for the turbine were off the shelf, I ended up tinkering with the machine intimately. Since no one else would climb the tower, keeping the machine running became my job. I ended up knowing the turbine better than the engineers," Cashman said.

When he moved back to California in 1977 to take a position at a Zen Buddhist Center in Marin County, he continued to be fascinated by harnessing wind power. Using the knowledge he had gained back in Massachusetts and Ottawa, he constructed a large, colorful windmill with cloth sails. A few years later, Cashman got a job in the Brown administration's Office of Appropriate Technology. Though he was initially put in charge of a composting program, Cashman was given wide leeway in determining what his ultimate job duties would be. When his boss, Robert Judd, executive director of the Office of Appropriate Technology, checked in on him one day, Cashman admitted to him, "I don't know the ropes around here." Judd replied, "Make your own ropes." Cashman decided that perhaps he should focus on wind energy. So he set up his desk, a hand-me-down from army surplus, right in the Governor's Office and without asking anyone just began working on wind energy policy.

"I wanted to break the vicious cycle. There was no wind technology because there was no capital. And there was no capital because there was no wind technology," said Cashman. "I was fundamentally an anarchist and government programs alienated me. They violated my sense of how the world really works." One of Cashman's first goals was to sift through policies that could accelerate development of wind power.

Cashman didn't think too highly of tax credits at the outset. And everyone else he talked to within the Brown administration was skeptical. It was too soon, too messy, they told him. However, two things convinced him it was the way to go. First was what he considered to be pathetic attempts by the federal government to develop smaller wind turbines at the Rocky Flats test site, the Colorado site of some of the nuclear industry's most gruesome fatalities. Average wind speeds there were a meager 3 mph. "What that means is that most of the year you get nothing. And then every once in a while you get a real howler, which of course tears apart all the machines," relates Cashman. These were the models, believe it or not, that the U.S. government was relying on for performance data in developing a national wind technology program.

Frustrated by his perception that the DOE approach lacked vision and meaningful assistance (that it was the ultimate big government approach to solving a problem), Cashman turned to history for inspiration in order to develop a unique California approach to promoting wind power. There he found a second reason to like tax credits. They had been employed by the English Crown to develop better longbows in the fourteenth century. He had just happened to be reading a book about how bow and arrow makers were exempted from taxes to induce production of superior equipment. While the credits were in place, a pool of highly skilled archers developed state-of-the-art bows that could shoot arrows over long distances. This approach proved to be far superior to knights in shining armor. "The sword and jousting poles of mounted knights were no match for peasant foot soldiers armed with the new technology of longbows," Cashman noted. In his mind, credits could not only change the course of battle techniques but also stimulate the development of a wind power industry. This history lesson confirmed in his mind that here was a government policy that could stimulate technological creativity. "We really weren't left-wing," Cashman pointed out, "My belief in self-reliance has more in common with the right-wing," he said, eyes reveling in the irony.

Still, in 1979 California was not yet on the national map when it came to wind power. That would change soon, as San Francisco hosted the first West Coast meeting of the American Wind Energy Association (AWEA). Up until that point, AWEA had been dominated by the "New England mafia," a small number of individuals with ties to MIT and the University of Massachusetts in Amherst, and a constituency of self-reliance buffs, led by Marcellus Jacobs, from the Midwest. Cashman had just ponied up the 25 bucks to join AWEA.

Jerry Brown was scheduled to give the keynote address. His backup was Wilson Clarke, who had authored a book entitled *Energy For Survival* and was Brown's point man in his administration on energy issues. On a whim, Cashman had decided to write his own speech. Then, as fate would have it, Cashman was given the green light to deliver the conference's main address when both Brown and Clarke failed to show up. "I explained my view of how technologies develop over time. California is where wind power plants will grow, I declared. We had concentrated winds that were excellent sites for wind power," is how Cashman summed up his speech. Though most in the audience had never heard of Cashman before the conference began, he was elected to the AWEA board the next day and at the next meeting of the board was elected president. "I rose to the top because I was the neutral figure. The Midwesterners hated the New Englanders, and so they were looking for somebody who could get along with both factions," Cashman said.

Later that year the California State Legislature approved Cashman's credits. In addition to the federal tax credits allowing a 25 percent deduction from federal income taxes, investors in wind power plants in California would receive another 25 percent deduction from state income tax for investments made between 1981 and 1985. John Bryson, the president of the California Public Utilities Commission, gave a major speech at the next wind conference, in Pennsylvania, extolling the new opportunities for wind developers in California. "Wall Street money people showed up and they seemed quite impressed. They had been turned off by the flaky solar guys," Cashman recalled. He was buoyed by the response from the financial community to the state tax credits, but he also received some disturbing news. The wind turbines on the drawing boards in California were assumed to operate over a 20-year period. A just-completed survey of machines in the field so far showed that the actual life span of the first generation of wind turbines averaged just seven hours. The tax credits would help get the machines in the ground, but could not ensure that, once installed, they would continue to operate.

That someone like Cashman was allowed to single-handedly develop such an approach to technology development could have occurred only with a governor like Jerry Brown. This appears to be an instance in which anarchy and government policy meshed. According to Robert Judd, Brown, "in his mysterious methods of indirect communication," had instructed him to encourage his staff to innovate. "Brown was a hands-off administrator. He wanted thinkers to experiment. The environment he helped create was not too unlike a technology incubator. Brown was willing to underwrite the cost of the greenhouse, to invest in new ideas. He was much more concerned about the integrity of the greenhouse than the specific plants to be grown in it. Whether wind energy or passive solar, Brown was genetically and generically supportive of an array of more sustainable technologies," he added. "Cashman fit the mold perfectly," he continued. "He had no preconceptions about the boundaries of what was appropriate. And Jerry Brown did not throttle or control people like Cashman. He expected something bright. If a project did well, Brown said he himself would take credit. If something went wrong, Brown would blame staff. Jerry Brown was not afraid of the risks that are associated with technology experimentation."

Brown's embracing of the "Small is Beautiful" values of the economist E.F. Schumacher was, in retrospect, most clearly articulated in his new wind power program. It was a breath of fresh air for the growing ranks of wind power entrepreneurs who thought that the federal government's priorities in its ambitious wind power commercialization program were all wrong. The familiar chorus of complaints consisted of one chief criticism. Most of the funds went to big, established companies that had little long-term interest in wind power; they squandered public monies that could have had a far larger impact if spread more evenly among upstart companies committed to a vision of a renewable energy future. California was nurturing soil to the new wind farmers not because it had superior winds, but because of policies crafted by Cashman that relied upon the greed of private investors to finance a new technology ballyhooed as a key solution to what was then deemed an urgent energy crisis.

The seeming schizophrenia of this approach should come as no surprise. California's legacy of exploitation of natural resources is epitomized by the legendary gold rush of 1849. Yet California is also ecotopia. A handful of northern Californians were developing small-scale renewable energy systems in the late 1970s. They relied on the plentiful sun, wind, or water to power their rural homes, which typically were not plugged into the grids of utility monopolies. With Governor Brown's blessing, these two diametrically opposed cultural camps—greedy exploitationists and well-meaning ecotopians—were brought together in a marriage of convenience that gave birth to today's modern wind farming industry. "Folks who follow tax credit industries are just like crooks—only they don't use guns," Randy Tinkerman observed. "What surprised me most was that the Wall Street guys became friends with developers like me, people they would have never dreamed about hanging out with before the wind power boom." Also added to the mix were military engineers with good postures, techno-geeks with lots of pens in their shirt pockets, and other renegades from various walks of life.

◆

California offered much more than just tax credits to add to federal credits for the wind farmers. It offered the kind of technical assistance that became irresistible lures to wind farmers from all over the world. It also offered a culture that embraced the dreams of restless entrepreneurs,

Bob Thomas helped deliver to the wind industry on both of these practical and less tangible fronts. With a wide, humble face and the body of an aging linebacker, Thomas doesn't look the part of a wind energy pioneer, but that shouldn't fool you. His interest in wind power dated back to the 1960s and grew as oil supplies became scarce in the early 1970s. At the time, he was working at China Lake, a U.S. Navy military outpost that, according to Thomas, encouraged creative thinking. "An individual here developed the sidewinder missile," he pointed out. Whereas other engineers at China Lake took an interest in nearby geothermal steam fields in southern California or in solar energy, Thomas was fascinated by the concept of generating electricity from the wind. Trained as an aeronautical engineer, he grew tired of testing weapons. "I wasn't quite sure if what the military was doing was really helping the country," he sighed.

Thomas had a recurring dream about a mysterious red six-pointed star. This dream began to preoccupy him and grow in significance in his mind. An avid devotee of Carl Jung, the Swiss psychologist who drew upon people's dreams to discover the meanings of their lives, Thomas tried to decipher the star-like symbol. Ultimately, he realized there could be only one meaning: It was a wind turbine design!

Three years later, at a seminar devoted to Jung's work, Thomas met George Wagner, the former president of the Jung Institute. Along with scads of movie stars, Thomas was also introduced to Sam Francis, a famous painter Wagner knew. Wagner had met Francis in 1972, shortly after the artist moved from Paris to Santa Monica. Wagner worked there as an environmental attorney helping to organize support, with the Sierra Club, to pass the California Coastal Protection Act. He had never heard of Francis, but Senator Alan Sieroty, a liberal Democrat with whom Wagner was working on coastal protection issues, informed Wagner that Francis was one of the biggest artists in the world. "His paintings were selling for a million bucks apiece," noted Wagner. Wagner was invited to Francis's house and he asked him to donate to that campaign. Francis gave a generous contribution. At the Jung seminar in 1975, the conversation drifted to the dream Thomas had. Francis showed some interest in the dream image of Thomas's and indicated he wanted to learn more about it.

Thomas took some micro-meteorology courses at the University of California–Davis and began to try to translate his mysterious dream into the nuts and bolts of a real-world piece of machinery. He constructed his first working model in his southern California garage. Though this crude model left a lot to be desired, "it was good to put something together," Thomas claimed. Shortly thereafter, Thomas met Sam Francis again, this time at a mutual friend's house in Los Angeles. Both Thomas and Francis were attending a seminar on the relationship between psychology and alchemy. Thomas's wife was an art writer and interviewed Francis for some articles about his art. Francis came by the garage one day shortly thereafter and quickly became fascinated by Thomas's concept of a new wind turbine whose fundamental design characteristics were based on the five points of a star. Francis surprised both Thomas and Wagner when he actually followed through on his promise in 1976 and wrote a check for $1.3 million. Wagner put up a quarter of his money and the Wind Harvest Company, Inc., was born. Thomas and Wagner, with the financial backing of Francis, began to chase a dream of merging the insights of Jungian psychology with modern engineering science in order to bring to life an entirely new kind of vertical-axis wind turbine.

Thomas, who has four kids, quit his U.S. Navy job and kissed his career civil service benefits good-bye. A fellow artist who knew Francis helped Thomas construct the next turbine model at the Camarillo Airport, where the Wind Harvest Company rented a hanger. "He was an artist, and a superb craftsman," remarked Thomas. "The turbine was completely hand-made." The model turbine was put on a trailer. Then a refined little test unit was installed near the Gorman Pass, a windswept region that straddles a small mountain range known as the "Grapevine," which separates the cooler, moist Los Angeles basin and the drier, hotter southern tip of the Central Valley near the oil wells of Bakersfield. The turbine clearly required more testing.

Though the dream had provided a general map of the design, dubbed "Windstar," transferring the dream image into a working wind turbine was proving more difficult than he'd imagined. Thomas began to panic about having given up his military-career safety net.

A casual acquaintance of Thomas's submitted Thomas's resume for a job he saw advertised at the California Energy Commission (CEC), a relatively new state agency charged with addressing environmental concerns about the state's energy policy. The California Legislature had passed a bill establishing a goal that 10 percent of California's electricity should come from wind power by the year 2000. Thomas thought little of this unsolicited submission. "I really didn't take the job prospect seriously," he acknowledged. Then he got a knock on his door. It was the head of the CEC wind energy program, Don Nichols. He wanted to offer Thomas a job. "I was getting tired of grubbing for money. And I needed a break from the pressure to come up with a working turbine," said Thomas. Given his family's financial needs, and this one-of-a-kind opportunity to learn about wind power, he accepted.

Within a week Nichols was fired, for reasons that remain unclear. Thomas suddenly headed up the whole state wind energy program.

At that point, the California wind power program was following in the footsteps of the federal government. The emphasis was on gigantic wind turbines, like the MOD-2 that was erected by PG&E. But Thomas, sharing the view of his fellow Jungian Cashman, thought the Brown administration should take a radically different approach. "We were all ready to take off, but we had no wind turbines. Instead of the big machines that the federal government was pushing, I wanted to focus more on smaller wind turbines," said Thomas. But there were few on the market. Federal research plans to develop "intermediate"-size wind turbines—the exact size that Thomas and Cashman wanted to see developed in California—were dropped.

The CEC conducted thirteen separate regional studies of wind speeds between 1980 and 1985, relying on a meteorologist traveling up and down the state in a helicopter, and other more traditional terrestrial techniques. At one and the same time the CEC thoroughly disproved the contention that California did not have sufficient wind to generate substantial amounts of electricity and identified the specific spots that had the most wind fuel. Providing wind farmers with these crucial siting data saved them the trouble of prospecting for wind fuel in a state the size of more than a few nations. All told, state researchers discovered California had the potential to develop 13,000 MW of wind power, enough to supply millions of homes. The CEC concluded that more than half that total potential supply was prime wind resource that could be cost-effectively developed right now.

To help realize this aggressive goal, Thomas organized California's first wind power conference with the purpose of assessing the current status of wind technology and the interest within the financial community. Palm Springs, which sits just at the periphery of landscapes where the Santa Ana winds, California's stiffest and most well-known wind regime, tear through the San Gorgonio Pass, was selected as the launching pad. People from all over the world were invited to attend the April 1981 event. Wagner remembers that the crowd was a far cry from what he expected or was used to on the Jung circuit. "Every conceivable shark in the country showed up. Guys who used to sell avocado groves, shopping malls, condos—you name it. There were all kinds of very expensive suits. It was like Oklahoma during the oil rush."

A 500 kW vertical-axis turbine manufactured by Alcoa was installed in a conspicuous spot as wind power enthusiasts from all over the world descended on Palm Springs. Thomas was excited that such a visible display of wind technology would be so close to the conference site. Since his design was also a vertical-axis machine (although one-twentieth the size), and since he had stopped all work on it while he was employed by the Brown administration, Thomas was curious about the Alcoa machine's performance. The wind power industry was so new that having some actual hardware in the ground would help create the impression that wind power was a real and viable business opportunity.

Just two days prior to the conference opening, the 200-foot-tall, three-bladed vertical-axis machine, whose rotors were curved like an eggbeater, created quite a stir. Vince Schwent, a CEC staffer monitoring wind power commercialization efforts, was there when it happened. After it had been operating for only two and a half hours, a computer glitch allowed the eggbeater to go into an overspeed condition. In other words, the brakes failed, and the speed of the swirling rotor was increasing with each revolution. "With a vertical-axis wind turbine you can't change the pitch of the blades to reduce stress on the machine like you can with a horizontal axis turbine," said Schwent. "When the brake burned out, the only thing you could do was watch from a safe bunker," he roared with an enthusiastic laugh. "It was great. The first long, skinny blade broke off at the bottom. Then it came off at the top and went striking out against the desert." The second blade cracked off at the top and soon the whole thing, including all the guy wires that had held the turbine in place, was a whirling mass of hardware. The whole huge structure collapsed with a gigantic explosion. It spun out of control and, like a small tornado in slow motion, ripped apart everything nearby. "If the accident would have happened the same day of the event, we would have all been killed," Wagner marveled.

The Alcoa turbine was billed as the main attraction at the wind power conference. Yet its dramatic death didn't deter the conferees, said Thomas. "You have to remember that none of these turbines had operated in the field for long. Blade failures were a generic problem with all of the early smaller wind turbines. All you could do was retrofit them and hope for the best."

Commercial development of Thomas's own Windstar was stalled because of his CEC job. He had signed a strict conflict-of-interest pledge when he accepted the job, so he missed out on the whole California gold rush of projects spurred on by the tax credits cooked up by his pal Cashman. There were plenty of other takers, nonetheless.

Chapter 10

King of the Hill

When the going gets weird, the weird turn pro.
—Hunter S. Thompson

"This is the place," Alvin Duskin remembers barking into the phone. "We should move the whole company out here to the Altamont Pass. We've got a friendly utility, a large substation, roads, and some incredible wind." Duskin is reminiscing about a very important conversation he had with Stan Charren, chairman of the board at USW and the man clearly in control of the purse strings of America's first corporate wind farmer. Although taken aback by the enthusiasm in Duskin's voice, coming through loud and clear over the long-distance phone lines, Charren hesitated to jump at Duskin's urgent advice. Charren was never one to act precipitously or to put all his eggs in one basket. The company had already put some money into purchasing the wind rights for 3,000 acres in Livingston, Montana—where people shoot guns at doors after days of mind-numbing Chinook winds—and 17,000 acres in Medicine Bow, Wyoming, another desolate spot in the middle of nowhere. There were also sites in Washington and Oregon.

It was 1980 and Duskin had just been given the task of sizing up the prospects for USW to take advantage of the tax breaks the Brown administration had suddenly offered in California. Since Duskin himself had previously designed similar tax breaks at the national level, he was well aware of how these financial incentives could help attract new investment in wind power.

In a previous life, in the early and mid-1970s, Duskin had made waves in California by authoring two statewide ballot initiatives to stop the proliferation of nuclear power in the state. Though both failed, these measures spurred legislators to block further investment in nuclear power plants until a repository for radioactive waste was established. Duskin was also involved with anti-nuclear initiative campaigns in six other western states.

A former designer of women's clothes in the '60s, Duskin had became an accomplished fund-raiser, and put on a number of anti-nuclear benefit rock concerts featuring big-name acts. But he grew weary of this profession, too. "To get some of these musicians to play for free I had to accommodate them in other ways," said Duskin, dressed in a collarless white shirt, with the demeanor of someone who likes not only fine garments but also really fine wine. "Too much of my work revolved around fulfilling demands for cocaine." By 1977, he was burned out on dealing with flaky rock stars and other celebrities and all the other stresses of uphill political campaigns and big stadium rock 'n' roll shows. Well-connected in liberal political circles because of his anti-nuclear work, Duskin went on to Washington, DC, to escape his self-assigned task of writing a book about nuclear proliferation. He ended up working for Senator James Abourezk from South Dakota. Abourezk sat on the 11-member senate energy committee, whose chairman was Henry "Scoop" Jackson, from Washington, a state that was hosting one of the nation's most aggressive nuclear power programs.

Duskin decided to write, and persuaded Jackson to introduce, a bill to take advantage of what he knew would be for him a short stint in our nation's capital. Instead of relying on federal government purchases, a traditional approach to jump-starting new power generation technologies, the new measure relied on private investors to help underwrite the development of wind power. His legislation became an amendment to the Crude Oil Windfall Profit Tax Bill of 1979. To an existing 10 percent investment tax credit, this new tax break added a 15 percent tax credit for wind projects.

What these tax credits did was make a wind turbine investment analogous to a real estate deal. Just think of the wind turbine as an income-producing building on a piece of land, with one key difference. The capital account that served as the basis for the tax write-off included not only what the investor had sunk into the project but also what the lender had loaned the investor. Typically, an investor would sink enough cash to cover a quarter of the wind turbine's cost. A common deal was a $25,000 investment in a $100,000 share of a wind turbine. The individual investor would get a loan to cover the balance. What the 25 percent tax credit translated into for the investor was an immediate $25,000 tax break. In addition, substantial tax breaks, totaling up to 75 to 90 percent of the original loan amount, would persist over seven years as depreciation was exhausted. "In essence, the investor [got] all of his profits for nothing," Duskin said. The "profit" was whatever income was sheltered from the IRS. Thanks to the tax breaks and depreciation schedule, it was as if wind farm investors never put up a dollar of their own money in the first place, and as if they never borrowed from the bank, but got to deduct from their income virtually all of the $100,000. "Of course, at the time we thought the wind turbines would all work." He lets loose a loud laugh, noting that revenues to investors—and more importantly to USW—flowed from actual energy production. "The tax breaks made wind very attractive. The investor got a lot of potential value for basically zero risk."

Duskin stated that all of the initial financing for USW came from wealthy friends or their acquaintances. Venture capital firms approached by Duskin turned up their noses. "They wanted to give us a small amount of money, see how it went, then keep adding money to it. We wanted to raise all the money we needed to proceed, to get out of the fundraising business and into the windmill business. So we looked for people who had developed companies from nothing to something large and successful." The first deal Duskin arranged with the use of his federal tax credits was the Crotched Mountain wind farm in New Hampshire. It was a $1 million deal and, like USW's new Altamont wind farms, was structured as a limited partnership. In this arrangement, a subsidiary of USW would act as the general partner while the investor donned the hat of the limited partner. USW would build the wind turbines and erect the wind farm for the investor; it would then enter into an agreement with the limited partner to operate and maintain the facility for an annual fee. The limited partner investor could continue to write off depreciation, maintenance, and the incidental costs of running the facility. For example, the Crotched Mountain installation generated just a smidgen of electricity on December 31, 1980. The amount of power actually generated didn't really matter to a wealthy San Francisco insurance executive by the name of Karl Bach. As limited partner investors, he and his two sons netted a $250,000 tax write-off thanks to the federal tax credits Duskin had engineered in Congress. They would have received a much larger tax break if they had made their investments in California because of its additional, equally generous state tax credits.

The proximate reason why Duskin had returned to California now, in 1980, and had been staying in contact with Charren by phone, was that he had been able to finagle a deal whereby the state's Department of Water Resources (DWR) would buy 100 MW of wind power from USW at Pacheco Pass, a site just south of the Altamont that researchers from Oregon had identified as a prime wind farm site. The state had leased 5,600 acres to Duskin for a project that would require 200 of the company's 50 kW turbines

Charren was happy that the unprecedented DWR power purchase contract was attracting potential investment capital from big institutions, the kind of financial backing that gave Duskin's efforts credibility with PG&E and officials within the Brown administration. But Duskin's call touting the Altamont to

Charren was prompted by what a state meteorologist had just told him. The wind blowing at the Pacheco site was about half what had been previously estimated. But the meteorologist also reported that a large parcel of land called Site 300, located on some of the highest ridges of the Altamont Pass region, was a far better site, if not the best site in all of California. That's when Duskin quickly got back in his car and headed north toward the Altamont, foot hard on the accelerator. The only data he could find about wind speeds for Site 300 came from the landowner—Lawrence Livermore National Laboratory, a Department of Energy facility responsible for nuclear weapons design and testing for the Department of Defense under the tutelage of Edward Teller. Site 300 had been used as a non-nuclear explosives test site in the 1950s. "The winds had been measured to make sure the explosives fumes didn't kill people in Tracy," said Duskin. They appeared to be quite wild. He claims to have snuck onto the off-limits site and flown a kite with a gauge attached that recorded wind speeds as high as 70 mph.

It just so happened that about this time, Bob Thomas of the California Energy Commission was just putting the finishing touches on an assessment of the wind resource at various points in the Altamont. After conducting his own reconnaissance at the Altamont, Duskin was in the CEC's Sacramento office pleading with officials to give him an early draft of the Altamont data. As he was leaving, having had no success, Duskin stumbled on an acquaintance, who happened to have said draft report in hand. Duskin asked if he could just take a quick peek at its findings. He then asked where the bathroom was. Right next to the men's room was a copy machine.

Duskin copied the whole thing and then found a phone booth across the street, where he made the fateful call urging Charren to focus exclusively on one wind farm site. Duskin had to threaten to quit before Charren relented and agreed to gamble on the Altamont.

Donning bib overalls so as not to look too slick to the Altamont ranchers, Duskin immediately went to work signing up roughly half the land and all the highest windy ridges identified in the pilfered report, for a mere $80! Indeed, the rights to the wind on the Mulqueeney Ranch—some 5,000 acres—were reserved for 10 bucks. In comparison, a 5,000 acre parcel obtained in Washington had cost USW $5,000.

Duskin's strategy was to price the rights to wind fuel along the same lines as contract prices for oil rights. Early oil prospectors used leases that tied up the land for a very small fee and in exchange offered the landowners far more lucrative royalty sums based on actual production. For a rancher in the Altamont, the real money would start flowing once wind farms were installed and started producing electricity. Duskin's most successful lure was a $100,000 payment once the bulldozers showed up to start construction. With Duskin's approach, USW did not have to pay this large fee up front. Once the project was financed with the revenue from tax shelters, USW could afford to pay ranchers a large lump sum.

With the help of PG&E, USW achieved a series of milestones in 1981. In March, USW's top brass met with PG&E officials to describe all the wonderful work that New England Electric Service (NEES), had done to get USW's machines into the New Hampshire soil, at Crotched Mountain, before the December 31, 1980, federal tax credit deadline. USW asked PG&E for similar help with designing and purchasing the proper electrical equipment and with construction. The biggest concern of the wind company was delivery times for new transformers. "We may have enough spare transformers to get the machines on line without waiting for normal delivery times," reads a March 13th PG&E memo. "USW could contract with PG&E for this help, as they did with NEES." PG&E agreed to cooperate shortly thereafter. This kind of cooperation was unheard of in the juvenile wind farming industry, where most wind farmers still viewed themselves as renegades on the frontier, a new breed of energy provider free of the trappings of the good-old-boy corporate utility networks.

On April 15th USW received zoning approval for its first Altamont Pass wind farm. "It makes sense to put windmills where the wind blows so hard that no one lives there," USW's Moore told the Alameda County Board of Supervisors just before they voted for the project. Because Moore, USW's first president, did such a good sales job in raising over $5 million with his "red book," he had come out to California to deliver the company's pitch in public forums. Duskin was there too, convincingly pointing out that the wind blew hardest in the Altamont Pass on the hottest days, when utilities such as PG&E had to burn more fossil fuels to respond to the energy-sucking air conditioners that customers turned on in the late afternoons and early evenings. USW then signed its first contract with PG&E on April 30th; construction of a wind farm consisting of 200 of USW's 50 kW wind turbines began at the end of July. On December 31, 1981, the first USW wind-generated electricity surged into the PG&E grid.

◆

In recognition of the virtues of AC transmission, PG&E had named a major substation on the eastern fringe of the Altamont Pass after Nikola Tesla. Construction of the Tesla substation began after World War II, in 1947, about the same time the last of the wind-powered DC renegades were plugging into Edison and Tesla's AC grid. The site now ties together power from 38 buzzing transmission lines. It is more than a little ironic that the substation that is the heart of the transmission system of what once was the largest combined electric and gas utility monopoly in the country—PG&E—would be named after Tesla. But in a sense, it is quite appropriate: The coming together of so many AC transmission lines in one spot has created a huge monument to high-voltage power.

Randy Tinkerman was the first to show me the Tesla substation, on our tour of the Altamont Pass. While standing on a ridgeline with an incredible panoramic view, he pointed out that the Altamont Pass was a naked landscape where the infrastructure of California's electricity, water, and waste are all clustered and exposed: The largest water canal in the world, the California Aqueduct, cuts through the eastern boundary of the Altamont, and a recent expansion of an existing solid-waste landfill, run by Waste Management, would make the Altamont the site of the third largest dump in the country. Tesla, the aqueduct, the landfill, a reservoir, and wind farms all create an industrial montage in a rural setting that speaks to how this landscape has often been viewed as someplace to be sacrificed for the greater good.

Wind farms operating in the Altamont feed into Tesla, adding juice to that coming down from the huge powerhouses in the Columbia River basin in the Pacific Northwest, and the coal and nuclear power plants scattered throughout the Rocky Mountains. Electric power is nondiscriminating. All electrons, regardless of fuel origins, congregate at Tesla and are fed into the grid. The Tesla substation became a focal point of wind farmers because PG&E proclaimed it would buy electricity only from these flaky wind farmers if it was delivered to the Tesla substation.

From John Eckland's point of view, the key to success in the Altamont Pass that first year was the Tesla substation. Eckland had just targeted the Altamont for his own wind farming operation. However, the wind rights to the land surrounding Tesla had already been licensed by Duskin for a pittance. So, Bud Steers, Eckland's real estate agent, worked hard to obtain an exclusive 50-foot easement on land surrounding Tesla to create a barrier around the substation. Eckland summed up his land buying strategy: "Nobody was going to cross us." Steers, nicknamed "the Cowboy," remembers what sealed the deal with landowner Mulqueeney: two half-gallons of Canadian Club whiskey. "I wrote the easement up right on the spot just off the top of my head. I told John that it was the most valuable document I would ever give him."

Though the immediate impact of the easement was that Eckland was able to wring unspecified financial concessions from USW, it was the long term strategic implications of controlling access to Tesla that drove their thinking. Steers is still proud of this maneuver. "We were driving around in these cars that were pieces of junk, operating out of our trunks, signing up about 10,000 acres of land for wind farms. It was what America is all about. We were hooked up to the stars and beyond. We were going to be rich beyond our wildest dreams. It was one helluva kick in the ass." Steers noted that he and Eckland would refer to themselves "as the 49ers of 1981;" the number "4981" was displayed on all of the company's gate locks.

Eckland was born and raised on a walnut farm in Stockton, yet he knew nothing of the wind farming legacy of his "City of Windmills" birthplace. Eckland became obsessed with wind power toward the tail end of a 15-year career at the Central Intelligence Agency. Eckland was the agency's top man at the CIA's Petroleum Supply Analysis Center. If anyone had his finger on the pulse of the world's long-term energy supplies, it would have been Eckland. What drove his initial interest in alternative sources of power was his concern about world stability that arose from the oil crisis of the mid-'70s. He was the chief author of a secret CIA oil market analysis that showed prices of oil rising to more than $100 per barrel by the early '80s. It was this alarming forecast that convinced President Jimmy Carter to don a sweater and declare "the moral equivalent of war" on America's addiction to fossil fuel energy sources.

Eckland declared commerce. He purchased the Fayette Corporation, a start-up wind turbine company located in Pennsylvania, in 1977, shortly after he left the CIA. Though he developed a patent on a solar energy device, Eckland soon shifted gears to focus almost exclusively on wind power: "Wind looked the most promising of all the renewables. It was the closest to being technically and economically feasible." His eyes still light up when he talks about his first major milestone in the Altamont Pass. On April 9, 1981, he and the Jess family signed on the dotted line. Eckland had purchased the legal rights to 600 acres. According to the contract, a percentage of revenues from the wind farming would go to Jess. This was a godsend to folks like Jess, constantly struggling to survive against the harsh Altamont environs. "Seemed like a pretty good deal for land that was too windy for subdivisions," notes Eckland.

Calculating the economic value of the haphazard Altamont winds, to be harvested by a technology with no track record, was hardly precise. Thus the earliest rates and terms varied widely in California. A typical wind rights contract in those years called for a flat fee of $500 to $1,000 per turbine, plus 2 to 5 percent of profits in the first few years and 10 percent in later years. Today, a standard wind rights royalty rate to landowners is 2 percent of the gross profit. This revenue stream has saved more than one family farm and ranching outfit—provided the wind turbines actually generated electricity. One large rancher in the Altamont realized as much as $400,000 per year in royalties.

Eckland set up a sales office right in the Altamont Pass near the Jess ranch, targeting mom-and-pop investors who were looking for tax shelters. Anybody driving along 580 could literally follow the signs to his office and Eckland would sell them shares in wind machines as if they were homegrown tomatoes. Eckland claims Fayette put up the first modern wind turbine in the Altamont Pass, the one painted red, white, and blue that still sometimes spins on the southern side of Highway 580. "We put it up in May on Joe Jess's property for his own electricity consumption. It was a very simple single phase machine," Eckland recalled. A dozen or so of the same turbine design had already been installed in New Jersey and Massachusetts, but Eckland was working on improving the performance of the machine in order to transmit greater amounts of electricity into the utility grid.

Since his manufacturing facilities were still based in Pennsylvania, Eckland had to wait impatiently for shipments of his improved design. The machines featured a larger blade, spun at a higher rpm, and were rated at 75 kW. Once the machines were on their way to California, the 20 or more plant employees also came out to the Altamont to install as many of the machines as possible before the tax credit deadline at year's end. "We didn't sleep for days," said Eckland, who did a little bit of everything, including pacing back and forth worrying about all of the little details that go along with installing wind machines and then connecting them to the all important grid in order to ship the electricity to Tesla.

It was an extremely wet winter that year, which meant that it poured and poured throughout November and December. The adobe soils in the Altamont "turned into snot," in the words of Eckland, and vehicles and cranes were sliding all over the hills during installations.

Eckland recalled from that first year in the Altamont a guy named Smokey, a pseudo-electrician who acted, more or less, like a supervisor. The only problem was that nobody got along with him. He so infuriated the two women working the field, Annie and Evon Castillou, that one day they filled his boots with cow shit. After the incident, he left, leaving the wind power industry for good. Eckland also enjoyed recalling what occurred on Mulqueeney's property as the Fayette crew struggled to build a distribution line to connect its turbines to the Tesla substation. The crew dug a row of holes about three feet wide. When they arrived the next morning to put in the wood poles for carrying the wires, they found one of Jess's prize bulls head down in a hole! It was so heavy that no one could figure out how to get it out.

None of the wind turbines Fayette planted along the Altamont's ridgelines to beat the December 31 tax deadline operated for more than about six weeks before being shut down for repairs (or blown to smithereens). Eckland was unfazed. "We're on the forefront of a new industry," he said in those early days when pitching his wind farms to potential investors. "I think we are making technology history."

Eckland attracted considerable attention with his Altamont wind farm. Among the foreign visitors to his farm were representatives from Taiwan, China, Greece, Germany, Ireland, Italy, Pakistan, Korea, and the Philippines. But most of his investment dollars continued to flow from small investors, many of whom saw the strange whirling contraptions as a chance to get something for nothing. "I think we got over 40 turbines in the ground that first year. Three of those were installed on New Year's Day, but nobody noticed. We got credit for those installations, too."

Chapter 11

The Tinker Man

The machine does not isolate man from the great problems of nature, but plunges him more deeply into them.

—Antoine de Saint-Exupéry

The Altamont Pass was not the only stage on which the wind farmers battled each other and the elements to cash in on California's new wind rush. New Year's Eve 1981 was a major milestone for wind farmers all over California. The California Energy Commission studies had identified several other prime spots where the winds were strong enough for wind farmers to do their thing. If USW represented the infiltration of the wind farming industry by New England corporate culture, many of the early wind farmers nevertheless shared a penchant for tinkering and small family business ventures, a tradition in the industry that was hard to kill.

Terry Mehrkam was one such pioneer. Checking out his first large-scale wind farm the morning of December 30th, Mehrkam wanted to make sure the turbines were energized and spinning away, converting the wind into dollars for the investors in his design. He also wanted to see how they were performing. While southern California can be nice and warm, storms were blowing in from the Pacific Ocean on this winter day, barreling down on Mehrkam's beloved wind catchers in Boulevard, located about an hour's drive from the San Diego coast and some 45 miles from the Mexican border. The wind tore through the desert like scissors. A strong wind is good for power production, wind farmers will tell you. But erratic, violent rips like these could wreck precious hardware.

Mehrkam had good reason to worry. The 60 or so wind catchers he had manufactured and installed all across the country had one little annoying tendency:

Every once in a while the blades would break off and fly away. They had done this from time to time back at the family farm, which sat next to an old red brick schoolhouse where his children played, but they hadn't so much as harmed a chicken. Nonetheless, Mehrkam was concerned that such random incidents could limit business opportunities here in California. A gutsy and committed promoter of wind energy, Mehrkam was not about to let a few "blade throws" ruin the opportunity that lay before him.

Mehrkam had been experimenting with wind turbines since 1973, but had never been able to keep them running cheaply enough to make decent money from them, at least not back in Hamburg, Pennsylvania, where he first began toying with them. Mehrkam had always been able to devise gadgets to solve virtually any problem. Before he formed his wind turbine company, Mehrkam was a mechanic, a welder, an electrician, and a metal plater. "You might say machinery is in my blood," he once said. At the age of 12, he designed and built an electric generating set. The son of a famous metallurgist, Mehrkam took courses at the Newark College of Engineering. He never got the degree because he wasn't the most disciplined of sorts. He'd originally designed and built his wind turbine at home as a protest against his local electric utility monopoly. When asked what he did, the quiet and reserved Mehrkam would simply say, "I am a tinkerer."

Wind farming in Pennsylvania was only a small-scale enterprise. Mehrkam expected that relying upon his own machinery and financing, he would see his investments in wind turbines pay for themselves in five years. His vision at first was no grander than to power his shop and home and to generate electricity for his brother and mother-in-law just down the street. But over time, he'd managed to turn his avocation into a comfortable living, owning his own plane to transport himself all over the country to service machines in Colorado, New Jersey, New York, Michigan, and now California. This despite the fact that his largest turbine, which was supposed to generate enough electricity for 50 homes, operated only 10 minutes before chewing up a gear case. Like every wind farmer, Mehrkam learned quickly that the wind is an unpredictable beast. "It's not the regular winds you have to worry about," he said after the ultra-thin 25-foot blades on his first turbine were sheared off because they could not take the stress that swells from velocity and torque. Though his first machine could withstand the run-of-the-mill prevailing winds, one still had to worry about the fundamental chaos that is wind. "It's the gusts—sudden bursts of wind. The wind is alive!" he'd exclaim. A Mehrkam turbine in Wyoming toppled over on a truck at a rest stop along the interstate. Another had been installed on Tom Hayden and Jane Fonda's ranch in Santa Barbara. Prominently located right next to a children's camp, it was operating without a hitch.

Mehrkam thought his time had finally come. In fact, a small flock of wind farmers descended upon the golden hills of California after Governor Jerry Brown rolled out the carpet for them with an unprecedented package of state investment incentives. Instead of lining up along streams, swirling their pans, itching for the shine of gold, folks were putting their fingers up in the air at sites on the windiest hills and ridges, seeking to strike it rich with an invisible bounty. Mehrkam, against the wishes of his wife and extended family, joined them.

The winds Mehrkam was trying to reap, and on which he had staked a substantial part of his future, were particularly flush. This, coupled with a set of tax incentives that virtually guaranteed success, was what made wind farming so appealing. There was, however, a catch. Wind turbines had to deliver energy to the utility before midnight on New Year's Eve in order for the wind farmer to qualify for both federal and state tax benefits for the entire previous year. Wind farmers sometimes bribed utility field workers with whiskey—or paid them under-the-table bonuses—to get them to hook up their wind turbines to the grid. One group of investors in the Altamont Pass were so desperate to get the tax breaks that they put up fake wooden blades on top of a phony tower to fool PG&E.

Mehrkam had just installed 30 of his four-bladed wind turbines in Boulevard, in the southernmost prime wind resource area identified by the CEC. December gusts that day in the California desert reached 50 mph or more. The wind catchers were supposed to shut off when winds averaged 25 mph, but something apparently went wrong. A machine went into overdrive: the props sped faster and faster until they spun out of control and broke off. Early wind pioneers employed a variety of techniques to stop runaway wind machines. Ropes were used to lasso blades. Once, to stop a turbine that had spun out of control for over 48 hours, an employee ready to leave the job site on a Friday evening used a shotgun. When he was finished, the poor contraption looked like a palm tree.

Mehrkam hadn't yet figured out an easy way to make his turbines stop spinning in high winds, but he had his own, peculiar braking technique. In the pouring rain, he quickly ran to the nearby tower (formerly a section of natural gas pipeline) and began climbing. As he had done many times before (and against everyone's advice), Mehrkam was going to stop the rotating blades manually. Brandishing a large screwdriver he always seemed to carry around, he climbed up the tower with lightning speed (and without a safety harness). Mehrkam must have felt the wind tear at his clothes. As he neared the top of the 40-foot tower, the vibrations must have shaken his body as if he were hanging on to a huge sledgehammer. Once he got to the top, he would have opened the clamshell-like nacelle, looked inside at the guts of the machine, and then stabbed his screwdriver into the brake. The sudden stop jolted Mehrkam. The clamshell cover closed shut, knocking him off balance. He fell to the ground.

Chuck Davenport, Mehrkam's CPA, was driving through the fog and miserable winter weather to talk with him about other planned projects for his clients. He arrived on the scene shortly after the tragic accident. "I pull up in my car, and there's Terry on the ground, all covered up," Davenport recalled. "Needless to say, it wasn't a pretty sight. To have one of the pioneers [in the industry] killed by one of his own inventions wasn't a great way to end 1981." Blades from his wind catchers, witnesses said, had broken off and zoomed toward a road. Davenport decided to comb the surrounding area to find the blades. "I found the first three I was looking for rather easily. I just walked a straight line and found them one after the other. Obviously, the air foils of the blades were well-designed because they traveled so straight. But I had to get a compass, and cross the road, to find the last one. It must have flown over 1,000 feet."

News of Mehrkam's death spread quickly throughout the infant wind industry, causing a few folks to have some second thoughts about their budding careers as wind farmers. Mehrkam's passing was a psychological blow. He actually had a company that was making money on wind power—even before America's last great tax shelter made wind farms the darling of Wall Street and of those who make their living counting and hiding other people's money. Paul Gipe, a long-time wind power proponent, was the first to write about Mehrkam's contrivances. A fellow Pennsylvanian and an author of many books about wind power, he also ended up writing the obituary. "He was like a Greek tragedy," Gipe said, claiming that Mehrkam's disregard for safety and the virtues of engineering science spelled his doom. "He was the classic backyard tinkerer," Gipe added, noting that Mehrkam bragged he could build the cheapest wind turbines by just using off-the-shelf parts available at any hardware store. Mehrkam also possessed an Old World sensibility. He belonged to a small religious sect whose members, like the Amish and Mennonites, were very suspicious of modern technology. "He was very ethical and he seemed to be out of his element as factors such as proper engineering, political forces and greed started entering the picture. He was among the old wind energy people who were left behind," said Tyrone Cashman, whose tax credits had brought Mehrkam to California.

Those who worked with Mehrkam used words like "genius" and "brilliant engineer" to describe him. Others were not surprised. when they heard of his death. "Mehrkam bit off a little more than he could chew," claimed Tom Gray, who succeeded Cashman as the head of the American Wind Energy Association in 1981. "I was always worrying that one of his wind turbines was going to kill somebody." Four black ribbons were tied to each blade of his first turbine on the Mehrkam farm on New Year's Day, 1982. To this day, he remains a Hamburg hometown hero.

Chapter 12

Good Connections

The usual trade and commerce is cheating all around by mutual consent.
—Francis Bacon

The arrival of USW in the Altamont Pass in 1981 marked the beginning of a phenomenal period of growth for the company. As Alvin Duskin promised, USW's big move into California had brought it friends in the utility business as well as in investment banking and other high places. The cozy partnership between USW and PG&E is the most revealing sign of how Duskin, Charren, and the rest were forging their own path through the business details of wind farming.

Jimmy Carter's National Energy Act of 1978, which established regulations and incentives based on Eckland's CIA forecasts of oil shortages and sharply escalating prices, had set the stage for the federal tax credits that Duskin helped push into law. It included other provisions to boost wind power. The Public Utility Regulatory Policy Act (PURPA), a companion measure to the 1978 act, required utilities to purchase electricity from anyone who generated it, at a price determined by the utility's "avoided cost." That is, the utility had to pay what it otherwise would have paid to build new power plants to meet consumer demand for electricity. PURPA was passed to create alternative power sources to the large coal and nuclear power plants already on the drawing boards of utilities. To implement this federal law, each state was to draw up its own regulations setting the terms and conditions by which independent private companies would sell their alternative energy to utilities.

The California Public Utilities Commission, which was filled with Jerry Brown appointees, had to set up contracts for the purchase of wind power by

PG&E. These contracts were also critical to developing a long-term revenue stream to attract capital from banks for the financing of wind farms. These contracts between wind farmers selling their electricity and utility monopoly buyers tell a revealing story.

Merrill Lynch had made gestures about wanting to sell wind farm partnership deals to its clients to take advantage of the new federal tax credits Duskin had helped make law. Nonetheless, even with the tax credits, the wind farm deals were not sweet enough. After all, these machines were not yet a proven technology. Because the wind farmers were independent, and not supported by a captive ratepayer base as utility-owned power plants are, Merrill Lynch wanted to see a little more certainty when it came to power sales. The long-term revenue stream mattered most to institutional investors.

Industry insiders had recommended USW to Merrill Lynch as the best wind farming firm, primarily because the team that Charren put together boasted several former CEOs who seemed to be loaded with financial and management expertise. PG&E recognized that, if it was to purchase power from wind farmers, it would need a credible, reliable long-term partner in the industry. Therefore, in 1981 executives from USW and PG&E huddled at the Holiday Inn in Livermore to devise an arrangement that would secure a long-term cash flow from a technology with no track record. They emerged with a set of terms for contracts in which the utility would commit to purchase power from a wind farm for 20 to 30 years. These gave PG&E and USW a better set of lures to attract new capital from limited partner investors put together by Merrill Lynch at just the right time.

This negotiated deal between USW and PG&E on the structure of power purchase contracts was then handed off to John Bryson, president of the California Public Utilities Commission (CPUC)—and Duskin's next-door neighbor in San Francisco—for regulatory approval. Bryson signed off on these power purchase contracts, which became models for all future wind farming investment deals and ultimately evolved into "standard offers."

The industry suddenly had fiscal credibility, and USW was positioned to be the first to take advantage. The reason USW rose to the top was Merrill Lynch. By 1982, USW had 100 wind turbines in the ground. And soon, thanks to the credibility Merrill Lynch offered the firm, USW was attracting the attention of those institutional investors that originally balked at Duskin's initial offers. Merrill Lynch proceeded with three back-to-back partnerships of $100 million each for new USW wind farms in the Altamont Pass.

USW's assistant treasurer, power contracts manager, and director of meteorology were all freshly hired from PG&E, too. The utility had worked closely with USW as it continued to tinker with its wind turbine—and indeed reimbursed USW for costs it incurred while building the substations and distribution lines necessary to carry the wind-generated electricity to Tesla. "The biggest check that PG&E wrote them in 1984 was for 1 million dollars," reads one confidential memo written by a PG&E employee after attending a February 5, 1985, meeting. "They [USW] expect to get a couple of 2 and 3 million dollar checks from PG&E this summer."

Having partners like Merrill Lynch and PG&E gave USW access to institutional sources of funding and expertise that had for the most part previously eluded wind companies. The Bank of New England established a revolving line of credit of $60 million to support ongoing operations at USW. Between 1978 and 1984, USW also raised nearly $20 million in equity capital to support its growth and its research and development and continuing engineering efforts. This capital came from eight venture capital companies. Among them were First Chicago Capital Corporation, Hambrecht and Quist, and the company's single largest stockholder, the Hillman Companies.

USW revenues for 1983, 1984, and 1985 were $29, $50, and $90 million, respectively; net income was a $1 million loss, then profits of $2.8 million and $6 million, respectively. In five years, USW arranged for over $300 million in private investment through the limited partnerships first pioneered for the Crotched Mountain wind farm (whose turbines lasted less than a year).

And the investors were quite happy with the money that came their way via the wind. Allstate Venture Group, an investment arm of the insurance company, registered its largest capital gains ever when it netted $30 million on a $2 million investment.

◆

USW proved it could play the political game and use its connections to gain a financial edge. But it also proved extraordinarily adept at turning its advantage in operating capital into a crucial technological advantage.

Both USW and Fayette used a three-bladed downwind, undamped, free-yaw machine. In other words, the turbine rotor would move freely in response to shifts in wind direction and speed. (These features went against the proven success of not only America's water pumpers but also the first small-scale electric generators developed by Marcellus Jacobs.)

The main difference between the USW and Fayette turbines was that USW could change the pitch of its blades by signals from a microprocessor, while Eckland's blades were fixed pitch. USW's pitching advantage, nonetheless, was irrelevant during 1981; all the blades of the early Fayette and USW turbines were ripped to shreds by the Altamont winds.

Randy Tinkerman was there in the Altamont Pass the day the very first wind turbines from USW were shipped in from New England in 1980. "I remember the scene very distinctly," he remarked. "Woody Stoddard, the guy who designed the turbines, who had just been fired from USW, was right there with me. There were six crates. We were trespassing, but we didn't give a shit. We opened one of the crates up, and Woody asked me 'What's the first thing that comes to your mind when you look at these machines?' And I responded, 'They look spindly, like spiders —undersized.'"

But USW's advantage in operating capital enabled it to change technological horses more adeptly. USW started off with a 50 kW machine having a 56-foot rotor diameter (a design known as the 56-50). Most of Fayette's turbines were rated at 95 kW. By the middle of 1983, USW switched to a turbine that could produce twice as much power as its original turbine. The basic architecture of the second-generation USW turbine was nearly identical to that of the first generation—the rotor, for example, was the same size. The size and strength of parts such as the generator and drive shaft were scaled up to accommodate the larger energy output. These machines worked far better than the 56-50s, but their light weight still made them susceptible to catastrophic failure. USW compensated for the fragility of its wind turbines by increasing the sophistication of the computer microprocessors it used for control.

The fundamental mission of the computer microprocessors in USW headquarters in Livermore was to regulate the angle of the blades relative to the wind speed. Start-up speed was 12 mph. Once the wind hit that velocity and held it consistently, the microprocessor would release the blades from the full feather position (parallel to the wind direction) and turn them to the power position (perpendicular to the wind). The blades would keep the full power angle until the wind reached 29 mph. At that speed, the turbines produced electricity at 100 kW. In winds ranging from 29 to 44 mph, the microprocessor could continually adjust the pitch angle to keep output at 100 kW. When winds exceeded 44 mph (which occurs from 3 to 5 percent of the time) the stress on the entire machine reached unsustainable levels. The computer would instruct the wind farm operators to return the blades to full feather position and bring them to a complete stop. The result was that USW was able to increase the operational availability of its 100 kW turbines, and reduce their frequency of failure.

Though Eckland can boast that he put up the first wind machine, with a little help from his friends, its blades were quickly trashed by the unrelenting winds. When inspecting his wind farms he would often find blades hanging from towers like "crumpled birds." It would be the first of many reality checks. According to Eckland, for the first few years of development in the Altamont, "every turbine was obsolete before it was in the ground. All of the wind turbines had the same problem." The Fayette turbine used a steel spar to connect the metal blades to the rotor. This was one weak link in the design, one that was never fully corrected. Though other firms used different materials, none of the wind turbines installed in those early years could stand up to the power and velocity of the Altamont wind. Machines were insured for a 10-year life. They often lasted 10 days—or quite a bit less than that.

◆

Due to the efforts of first PG&E and USW, and then a woman named Jan Hamrin and the CPUC, wind farmers could choose from four different contract options to sell their power to the utility. USW chose a power purchase contract whereby energy prices were fixed at a set rate of increases for the first 10 years of wind farm operation and then reverted to market prices for energy from the eleventh year on. This came to be referred to as Standard Offer 4, or an "SO 4" contract. (At the time, oil prices were used as the proxy for energy prices. Because of changes in the fuels used to generate electricity in the 1980s, the price of natural gas would eventually replace oil in determining utility avoided costs.) Though companies that signed these contracts gave up some potential profit in the near term—and risked locking into prices during the first decade of operation below future oil prices—the fixed escalation in energy prices for the first 10 years was a guarantee attractive to all the money guys.

Most wind companies chose the SO 4 contract. Some certainty in the uncertain world of electricity was appealing.

Eckland's Fayette was the only major wind power company to choose a different one of the power purchase contracts that was also on the table. The SO 1 option directly tracked real energy prices, not the projected price increases, after only five years of operation (instead of after 10 years as with the SO 4 contract). No. 1 was potentially more profitable, but was also far more risky. Eckland's decision was based on a hunch. He had closely monitored the world's political climate and trends in international monetary and public policy while at the CIA. He was convinced that many of the projected price increases in oil and other fossil fuels were actually understating the costs of an energy crisis. Eckland believed the world was on the verge of cataclysmic change in its power supplies. He gambled that the energy crisis would only intensify in the very near term.

Duskin monitored his rival Eckland's strategy suspiciously. Duskin was convinced that Eckland had inside CIA information on oil prices and was being the coy fox, since he was virtually the only wind farmer to pick the SO 1 pricing. Duskin urged Charren to follow Eckland's lead and sign power purchase contracts that would follow actual oil prices—sooner rather than later. Duskin was outvoted.

Though payments to wind farmers were actually below utility avoided costs in the first couple of years of operation, the energy payments in the SO 4 contracts then quickly rose according to its escalation formula to go well above prices for oil. Since Eckland was still getting paid the same for energy as his competitors during the first five years of operation, he wasn't sweating yet about his gamble that oil prices would rise steeply over the next few years. Like USW, he kept raising funds because of tax shelters. But he was at a different scale, appealing to a less sophisticated, as well as arguably seedier, investment audience. His CIA connections didn't seem to really help him much in the energy business.

Nonetheless, Eckland's wind farming business grew, though slowly. The 40 turbines (representing 3.8 MW of total capacity) that had been installed by 1982 generated only $10,000 all year long from less than 100,000 kWh of electricity. But by 1983, a year when the Altamont Pass produced four times as much wind power as any other California wind farming region, Fayette came in second behind USW in energy production. In 1984, investors shared $2 million in returns. In 1985 Fayette promised a threefold increase in proceeds, projecting 91 million kWh of wind power production in that year. Nonetheless, 1985 proved to be disappointing; Fayette's wind farms generated only 51 million kWh. Fayette blamed the low production figures on a bad wind year. Just as in other forms of farming, folks in the wind biz sometimes enjoy a bumper crop but also suffer the consequences of a bummer one. Fayette claimed that available wind fuel in 1985 was 25 percent below the company's projections. But the turbines produced 44 percent below what was expected. Thus, the turbines' less than stellar performance was apparently liable for the additional 19 percent drop in energy production and related revenue losses.

By the end of the California wind rush in 1985, Fayette had 1,600 wind turbines in the Altamont Pass, representing 152 MW of electrical generating capacity. Those figures earned Fayette the distinction of being the state's number two wind farm developer in terms of the amount of potential electricity that could be generated by the number of turbines it had installed in the ground. Nonetheless, in 1985, SeaWest, with only one-fifth of Fayette's installed capacity, produced more wind-generated kilowatt hours of electricity. Of the 28 wind turbine designs installed in California by this time, Fayette's machines were rated sixteenth in actual field performance. The best machines at this point, Danish Micons, generated electricity 23 percent of the time. Fayette's machines worked only 5 percent of the time. (Most wind turbines operating today in the Altamont Pass register annual capacity averages in the high twenties or low thirties.) Poor siting was part of the problem, but the energy output of Fayette's turbines was also less than half of other turbine designs operating right next door.

By 1986, the formula in the California power purchase contracts dictated a price of roughly 10 cents per kWh. Unfortunately, Eckland didn't have to broadcast to his investors that he wasn't getting paid that amount after five years of operation. The lower payments were a consequence of his choosing an SO 1 utility power purchase contract. Oil prices were going down, not up. Eckland kept waiting for the next war or for members of OPEC to return to their price-gouging ways. Instead of the 10 cents per kWh that USW received in 1986 for its 1981 machines, Eckland's first-year machines were receiving payments in the 3 cents per kWh range. In addition, Eckland had not created a limited partnership to develop wind farms, as most wind companies had. Fayette was a simple proprietorship—like the small neighborhood grocery store. Thus investors never had complete access to the financial details of Eckland's undertakings.

Of course, once investors discovered that their turbine wasn't producing the electricity projected because of its location and problem-plagued technology, they were not too happy. They also found out that their $26,000 investment, made in order to get an annual tax write-off of some $45,000 for five years, really wasn't worth that much because the turbines were not producing the promised power.

This quandary on the part of investors, ironically, kept Fayette going for some time, since Eckland was able to continue to pay out profits to his early investors with funds he apparently raised from fresh investors. No one wanted to tattle on Eckland's production shortfalls, because, once the money was invested, owners of Fayette turbines risked the nose of the IRS sniffing into their finances and nullifying the tax savings they had claimed. One Fayette investor—Ralph Koldinger of Sacramento—called for a congressional hearing when the IRS did get involved, arguing that the IRS was handling its investigation of Fayette in a "capricious" manner and that well-meaning investors in Fayette's wind turbines were being used as "negotiating gambits in a game where we are unwilling players and in which we will lose even as we win."

◆

As the years rolled on, a clear victor emerged in the race between Eckland's Fayette and USW to capture the most coins from the winds that growled in the Altamont Pass. While neither's turbine design was close to being perfect, it became clear that one of the two companies looked a lot better than the other when all the production numbers were added up on paper.

Eckland admits today that his land-buying strategy backfired because it was too centered on the Tesla substation. Steers and Eckland were able to wring small financial concessions out of USW so the company could transmit power over land it had already leased, but the effect was inconsequential. Not only did PG&E relent and take power from other locations in the Altamont, but most of his turbines were sited on the lower ridges and were downwind from the USW machines. The thinness of the Altamont wind fuel came back to haunt him. "The USW turbines took one and a half miles per hour right off the top," Eckland lamented, referring to how turbines pulling energy out of the air can reduce the performance of machines downwind from them. Eckland failed to recognize the importance of the nature of wind at very specific points on the skin of the earth.

Eckland thinks another five years of tinkering would have produced reliable wind turbines. Before the IRS ruined his life by challenging each and every tax deduction made by investors in his company, he had purchased the largest blades on the market from the Danes, whose wind turbines held up the best in the Altamont and other wind farm areas. The firm began testing a 250 kW turbine in 1984, but its performance left something to be desired. An even larger machine, rated at 400 kW, was plagued by drivetrain failures and never really entered the commercial market.

It took me years to track Eckland down, and when I later tried to get a hold of him again, nobody seemed to know how to do so. I had heard so many terrible things about him, that I was a bit surprised when we met face to face. He appeared so humble, friendly—and forthcoming. Still, Eckland's story is one of the oddest chapters in the history of wind power. ("That guy belongs in jail," Tinkerman volunteers every time I mention Eckland's name.) At its peak, Fayette hired 200 people and was the focus of countless magazine and newspaper stories about how a guy from the CIA who was concerned about the energy crisis was trying to be save the world with wind power. There are always two sides to every story. The history of Fayette is a tale of deception, but it is also a tale of underdogs, perhaps altruistic in purpose, but who, operating on shoestring budgets, may have betrayed some of their initial ideals.

Chapter 13

The Pirate Poet of the Tehachapi

The courage of the poet is to keep ajar the door that leads into madness.
—Christopher Morley

"I was supposed to be in Hawaii."

That's what Jan Hamrin remembers about California, December 1981. Instead of black sand beaches and tropical rum drinks in the sun, she shivered in the snow in the Tehachapi Mountains, a stretch of young million-year-old rocks that separates the vast Mojave Desert from the massive human populations of Los Angeles. This is where the famous San Andreas fault crosses the lesser-known Garlock earthquake fault; the land moves one centimeter per year, which is a gallop in geologic time.

The Tehachapi Mountain range is far less notorious than the Altamont Pass, being in the middle of what most tourists that descend upon the Golden State would say is absolutely nowhere (by California standards, anyway). Of course, the Altamont Pass is hardly a well-known landmark, either. Nonetheless, the Stones show, its proximity to San Francisco, and the arrival of USW, always the most visible of the U.S. wind farming companies, made the Altamont Pass the focus of most early media coverage of California wind farming. Bakersfield, home to much of the state's oil and natural gas industry, and considered by many to be a redneck wasteland, is the closest "big" city to the Tehachapi range.

Instead of lulling in a romantic getaway, Hamrin ended up spending her vacation time with a wind farm crew that could have walked on site from a TV sitcom. Her boyfriend, Tim Rosenfeld, had just joined up with a company, Pacific Wind

& Solar, that was going to get bloody rich by installing a few wind turbines before the end of the year. Like Ty Cashman, Rosenfeld had worked on energy conservation and renewable energy issues in the Jerry Brown administration, but he tired of bureaucracy. "I was given an offer I couldn't refuse." An entertainment industry lawyer had just raised a million bucks for a wind farm that was to consist of 10 wind machines. But to reap the full benefits of the Cashman tax credits it needed to go in ASAP. Hamrin and Rosenfeld jumped in to install the wind farm even though there was no guarantee anybody would buy what little power it might actually produce. It didn't matter. Investors, including Hollywood celebrities, had money they needed to sink into shelters to shield it from the IRS. And wind farms, regardless of the real-world constraints, were the hot shelters this year.

There was one small problem. Nobody on the team had ever installed a wind turbine before. Pacific Wind & Solar had an awfully hard time recruiting experienced workers. They did, however, come across a motley motorcycle gang that just happened to be unemployed. It was the off-season for oil drilling work in nearby Bakersfield, and wind power seemed right up their alley. Bikers were valuable in the eyes of many wind farmers because they knew what to do with a torque wrench. It was Hamrin's job to open the trailer every morning and get the gang started on the job. She fondly remembers the leather-clad crew, the roaring Harleys. "There was this huge bear of a man. His name was 'Muffy.' I didn't ask any questions," she chuckled. "Muffy was great to have because he could lift things that nobody else could. There was also a guy in a ratty black parka with greasy fur, he sort of looked like a ferret. I remember him telling me there was something wrong with the fourth turbine. I asked: 'What makes you think that?' He didn't say, but it turned out he had an excellent eye. He was a mechanic and ended up getting a full-time job on the maintenance crew."

The crew was almost lost when a rival biker gang with even bigger and better Harleys, and much stronger feelings about the notion of territory, showed up at the site one day. Apparently, the Tehachapi was their turf. "If our crew didn't leave, there would be bloodshed," Hamrin recalled. After about 20 minutes of discussion, the biker work crew made a unanimous decision: burn the jackets and get rid of any trace of biker gang insignia! They cast their fate to the Tehachapi's winds. "It was really the wild, wild West," said Hamrin. The attitude was: "Yahoo! Let's do it! We all knew we were involved in something new and different and dangerous, yet here we were, in such picturesque surroundings." Sometimes the serenity would be disturbed, however, by Stealth helicopters and ground-hugging missiles from nearby U.S. Air Force bases. The military presence just added to the mystique of this odd job, added to the sense that history was in the making.

It turns out only one of the 10 turbines developed by a couple of engineers at the Department of Energy laboratory at Rocky Flats, Colorado, and installed that December by Pacific Wind & Solar actually produced electricity. But because the other nine turbines were part of the "farm" that produced power, all 10 turbines (the full $1 million investment) were eligible for the maximum amount of state and federal tax credits. These turbines were among the noisiest ever invented, the squealing and squeaking clearly audible over the large rushing sound of the wind.

◆

The first time I drove through the Tehachapi Mountains was in the dead of winter at dawn. Mist was rising everywhere from a fresh storm that had just dumped some very wet snow. In and out of the fog I could see the sleek white turbines spinning away, juxtaposed against dark green hills and occasional bright snow patches. The majority of wind turbines installed here since 1981 were imported from Denmark. Generally speaking, most look alike and perform well, two attributes that help make this wind farm area seem more in tune with its surroundings. Whereas the Altamont's random assortment of highly diverse and often faltering wind turbines conjures up an image of chaos, the wind farms here for the most part look like they actually belong here. The weather in Tehachapi is far more extreme than the relatively mild Mediterranean climate of the San Francisco Bay area. The Tehachapi mountain ridges explode in April and May with incredible displays of spring wildflowers. Orange, purple, red, and yellow blossoms turn these hillsides into a floral mosaic just as the winds start to pick up. The temperatures are higher in the summer and much, much lower in the winter than in the Altamont Pass. This is a part of California where people know what a snow shovel is. The weather is a primary reason few come to visit and even fewer stay.

The first settlers of the Tehachapi range were a small nomadic tribe of 500 to maybe 1,000 people. The Oak Creek Pass cuts through Tehachapi and is where most of today's wind turbines are planted. The Kawa iisu tribe, which roamed nearby ridges and hills, sometimes traveling as far as the Mojave Desert, discovered it about 3,000 years ago. A 1998 addition to the California State Park system features their Sand Canyon winter homes. The tribe would migrate to lower elevations as snow fell and food became scarce. They constructed their dwellings of bent-over willow saplings, creating what looked like upside down baskets. Not coincidentally, these people were among California's finest basket weavers. In the spring, they would disperse, typically in family units, and harvest acorns and pinion nuts, traveling far and wide. They would return to these homes the following winter. Once white settlers entered the picture, the tribe declined rapidly, not because of any campaign of genocide but from disease.

A series of visitors subsequently passed through the Oak Creek Pass, but few found what they were looking for. In 1776, a Franciscan friar named Garces came here looking to establish a mission. Most of California's 21 missions were located along the coast, on islands, or on big rivers up north. Tribes living further inland tended to be nomadic hunter-gatherers and were resistant to the words of Jesus Christ. Garces couldn't find enough potential converts to justify a mission. In 1803, Father Jose Valvaida showed up, searching for Native Americans who had escaped from the mission in Santa Barbara. He, too, was disappointed. The first English-speaking person to arrive on the scene was Jedediah Strong Smith, a fur trapper and scout for the U.S. Army. In 1834, he poked around these hills to see if there were enough beaver to trap to keep up with the demand for felt hats. The streams here weren't big enough to support beaver populations. He was attacked by Indians and barely escaped with his life. In the 1850s, Oak Creek Pass became a major travel route of the Butterfield Stage, which transported the growing horde of gold-rush prospectors pouring into California, to gold mines in Havilah, some 35 miles to the north. Then, by the 1870s, a drought in the San Joaquin Valley prompted shepherds to search for new grazing areas. They stumbled up the Oak Creek Pass and walked their sheep by foot through this howling stretch of land, and continued to do so every year until 1971.

◆

The first Tehachapi wind power company to successfully negotiate a power purchase deal with an electric utility was Zond Systems, Inc., which installed its first machines right next door to Pacific Wind & Solar's site in December 1981. Zond was founded by James Dehlsen in October of the previous year. Born in Guadalajara, Mexico, to a Danish businessman and engineer who worked for the Southern Pacific Railroad, Dehlsen is a life-long entrepreneur who started out, literally, on neither side of the tracks. "We lived on the line in a railroad car," said Dehlsen, who sports a black patch over his right eye. "That part of Mexico was wild back then. Everyone wore pistols and there were banditos all over the place." In contrast to USW and its New England roots and connections to old wealth, Dehlsen's story is one peppered with trials and tribulations that were overcome with the help of new California money.

Before entering the wind power business, Dehlsen had served in the U.S. Air Force, graduated with an engineering degree from San Diego State University, dropped out and gone sailing for a year, got married, and created an international stock investment corporation that raised $200 million in two years. Dehlsen returned to southern California in 1980 when the stock market soured, looking for new challenges. He didn't waste much time, inventing Triflon, a high-speed lubricant that revolutionized tribology (the study of friction) in this country. Among the biggest fans of Dehlsen's product were participants in the Human-Powered Speed Championships, who frequented alternative energy fairs. "In order to succeed in human-powered machine competition, you have to squeeze every last ounce of friction out of the moving parts," Dehlsen commented. "It's also a critical component of all alternative energy machines. That's where Triflon fit in, and that's what got me started on the renewable energy idea."

Dehlsen found religion in the Palm Springs wind energy conference in 1981. "By the second day I was convinced that wind was the way to go. All the independent studies on wind were tied together at that conference. All of its advantages were confirmed. It's clean and it's abundant. Wind turbines can also be installed in a fraction of the time it takes to build conventional power plants." Most importantly, Dehlsen saw that the investment community was supporting wind power because the government was offering great tax advantages to wind investors. "Arthur D. Little predicted that wind energy would become a multi-billion dollar business, and we decided right then to be the leaders in the field," he said.

Dehlsen and his wife quickly purchased 750 acres on a windswept ridge of the Tehachapi Mountains with the capital he had raised from his various entrepreneurial endeavors. Unlike the Altamont Pass, where one could lease large tracts of land with very little money, the Tehachapi wind resource area had already been subdivided into smaller plots. But the land was considered marginal even for grazing, so it was cheap. Dehlsen recognized it as a wind farmer's paradise. "We kept driving around looking and we knew we'd found the right spot because the wind was blowing so hard we couldn't get the car door open," Dehlsen said.

A neighbor, Daniel Reynolds, a designer of ocean-racing boats, and Bob Gates, an aviation and marine equipment salesman, were Dehlsen's two initial partners in Zond.

The three set a firm goal for their company in April 1981. Gates remembered it this way: "We stood up on the hill and promised each other that we would have 15 turbines on this piece of land by the end of the year, come hell or high water. We're going to make it happen. We're never going to take no for an answer. We're never going to say die. We're never going to listen to reason. We are just going to do it."

Because they did not want to associate themselves with some of the kooks in the industry, Zond originally didn't want to have to rely on the state and federal tax credits available to attract investment capital. Nonetheless, Gates, who grew calluses on his fingers dialing for potential investors, reluctantly endorsed going the tax credit route in order to live up to the commitment they had made.

The wind farm, dubbed Victory Garden, was approached from the perspective of a landscape design. They had rendered over 200 separate drawings, showing where the individual wind turbines would be placed: which ridgelines, the length of turbine strings, height and width, etc. Zond then searched high and low to find the best wind turbines for the site. Unlike USW, Zond did not yet have the capacity to design and manufacture its own. There were not a lot of choices, so they went with local designer Ed Salter's "Storm Master" turbines.

Salter vividly remembers meeting Dehlsen for the first time. "I was up on a tower putting up a test unit when a friend—Fred Carr—drove up, and this bookish lanky gentleman popped out of the car. His name was Dehlsen and he wanted to use my turbines for a wind farm, provided we doubled or even tripled their energy output." Carr ran an alternative energy business. Though it was primarily solar water heating installations, Carr and Dehlsen had installed a crude Australian DC windmill on an oil platform in the Santa Barbara channel.

The meeting between the three men led to an agreement by which Zond purchased over 450 machines. Zond eventually purchased the patent rights to the Storm Master, even though the turbines never really worked well. "It's taken me 15 years to get to a solid production unit," notes Salter today, pointing out that without large R&D budgets, most individual turbine designers proceeded on a slow learning curve. Though he sold the rights to the name and design, he has continued to work on it, and other turbine designs, to perfect performance because he, like others in the industry, is obsessed with making the machines work.

Interestingly enough, the progenitor of Storm Master design was again the German Ulrich Hutter. The concept was to feature slender fiberglass blades that could bend, like miniature vaulting poles, in response to hurricane-velocity gusts. Ironically, the bigger Storm Masters designed for Zond were a disaster, because they had not been tested—anywhere—before installation. On top of that, the gearboxes, manufactured in India, were so leaky that lubricant seeped through, splattering maintenance crews below. Eventually, the blades were taken off all the machines, trimmed by a third, and then welded to the hub. These modifications were only Band-Aids. All 15 eventually self-destructed under the constant abuse of the Tehachapi winds. Salter explained: "We hadn't done any development work on the 12- instead of 10-meter machine and then got caught in the revolving door of letting new sales finance a retrofit program. We didn't have the time for any proper R&D."

Lloyd Herzinger, company photographer and jack-of-all-trades, has an interesting story to tell about the Storm Master: "I was out on a bright sunny day and noticed the first machine installed up on the top of the hill was running, so I grabbed my camera. I took some shots and then ran down to the trailer to get warm. When I headed back up, I noticed that one of the blades had been thrown. The machine must have gone into overdrive. As I got closer to my camera and tripod, I noticed, because of the footprints in the snow, that the blade landed right where I had been standing. It would've gotten me."

"War Zone" is the way Herzinger remembers that first year, as well as subsequent years, planting wind farms in the Tehachapi. "Here you were out in this crisp, clean mountain air, but all around you: tons of mud, the smell of diesel trucks, the sound of cement trucks pouring." He used to hang out on the beaches of Santa Barbara making metal sculptures out of scrap. But this December, Herzinger was slaving away in blizzard.

Dehlsen told me he views the early days fondly. "It was almost like trying to manage under wartime conditions. What Dan and Bob were up against, what we all were, was really life and death. Because we already had committed to the infrastructure. It was already being done. We had a convoy of cement trucks going up there. We had 15 containers on the docks in Houston, two or three turbines a day coming on the site by December 1. I mean, we were committed. We brought in these gigantic floodlights so our crews could work at night. It was insane."

On December 15th, some 16 days before the New Year's Eve tax credit deadline, Zond got Southern California Edison to agree to purchase its wind-generated electricity. Six months of negotiations had actually yielded results. Zond had come away with what would become a model utility power purchase contract along the lines of what USW had negotiated with PG&E. Many other wind companies, including Pacific Wind & Solar, had been spending way too much time before state regulatory bodies trying to figure out the terms of the utility power purchase contracts. Like USW, Zond had taken a different course. It went directly to the entity that was purchasing the wind-generated electricity. It shook hands with an electric utility—commonly referred to as "Edison" throughout the West—that collected more money from its ratepayers than any other electric utility in the country. Not many wind farmers could do this. While not as well-financed as USW, Zond had considerably more clout than folks like Mehrkam or Pacific Wind & Solar.

Still, the power purchase contract was worth little if wind turbines did not get in the ground in time and start spinning to produce at least a few kilowatt hours of electricity before the end of the year. Gates recalled, "The crew was working seven days a week. One of their wives, pregnant, was fixing buckets full of hot stew and bringing them up to feed the men something warm so they didn't freeze and could keep on working. Christmas Eve—it was just a howling blizzard—they hook up the wires, release the first turbine. On come the lights! We were all just jumping up and down. It was marvelous." On New Year's Day, Dehlsen patrolled as snow flakes coated the scene, and he found workers passed out in cranes, exhausted from the grueling schedule of round-the-clock installation.

◆

The majority of the turbines now operating on the Dehlsen wind farm are the stout Danish turbines, to which Dehlsen ultimately turned after the Storm Master proved incapable of withstanding Tehachapi winds. These Vestas turbines are much heavier and sturdier than those of American make. One way to look at the differences between Danish and American design philosophies is that, whereas the Danes faced up to the severe challenges of capturing the power of the wind by building sturdy structures whose rotors were directed into the oncoming wind, Americans designers instead looked away, pinning their hopes on skinny turbines whose backs were to the wind, as if to ignore the harsh realities of the demands of weather on machines.

To prove that humans could co-exist with this new form of farming, Jim and Deanne Dehlsen decided to make the base of the Tehachapi wind farm home. The house built there became a showcase of energy and ecological efficacy. The Spanish-style villa is composed completely of recycled materials. Wood from wine barrels, old farmhouse boards, old railroad ties, scrap and pieces from neighbors, and other construction materials found throughout the local environment all were integrated into a structure that merges with the landscape. The household is totally dependent on local natural springs for water. It is a mix of cheery functionalism and dramatic, bold art. But as one gazes out of the windows of Dehlsen's home, a huge cement kiln fills the view. Its eerie lights and gigantic spherical shape transform this desolate stretch of Highway 58 into stark portraits of contrast: the whirling white wind turbines on one side of the road, the large dark sphere-like industrial infrastructure of the cement kiln on the other.

The image that stands out in my mind was from my second visit to Dehlsen's home, which just happened to fall on a full moon. I heard coyotes howling. The sound seemed to be coming from a few feet away from the cottage near Dehlsen's house where I was trying to sleep. As I peered out the window, I saw the shadow of one or two coyotes. The image etched into my imagination, however, is an army of wind turbines spinning in the bright lunar light. I felt like I was on another planet. You couldn't really hear them—the wind was too large—but could almost feel them. I hardly slept that night and repeatedly got up to peer out different windows, each offering a different perspective. I finally took a step outside. The wind grabbed me as I watched a few coyotes scatter into the hills now filled with a synthetic whirling motion. The panoramic view was wild, ridgelines flowing into one another, turbines disappearing or appearing to float among the sharp rocks as I walked around. I must have gazed at the surreal display, pacing back and forth, for a half hour or so before chilling and going back inside to finally fall asleep.

The Dehlsens resigned themselves to live here, apparently dedicated to demonstrating that their home was a satellite station on the frontiers of a revolution in power-generation technologies. They learned to live with, and from, the wind. Forfeiting their domestic privacy for the sake of PR, the Dehlsens allowed their ranch to become a popular hangout for anybody who was somebody in the wind industry. Typically, it would be an intimate evening. Neither

Dehlsen was bashful when it came to the consumption of wine. When everyone was a little drunk, Jim Dehlsen would invariably launch into his sales pitch on wind power.

He once took me to the top of the mountain in his huge weather-beaten Bronco. The ride up was an adventure, to say the least. It had recently rained and part of the road had washed away. After a series of swithchbacks, Dehlsen came to a sudden stop. As he peers up at the spinning wind turbines in Victory Garden, you can feel his intense pride. He squints, adjusting the black patch over his right eye and running his hand through light brown hair. "Wind energy just makes so much sense," he gestures with his hands at the panoramic view, motion from white propellers surrounding us on this crystal blue morning. As we get back into his groaning Bronco and careen about the wind farms surrounding his ranch home, there are breathtaking views. The giant blades of the Danish Vestas wind turbines seem to put both of us into an almost mystical trance.

According to Dehlsen, the site of the first wind farm here in the Tehachapi is very special: "It's something you can sense. It's almost magic. In fact, there's a lot of magic floating around this whole project. For instance, every one of the Zond team sails; they're addicted to wind and always have been." Why name the company "Zond?" Dehlsen just made it up off the top of his head. One night, after consuming more than a modest amount of barley beverages, Dehlsen searched the dictionary to see if any such word existed. Though "Zond" did not appear, the word "Zonda" did. The term referred to the winds that blow down from the Andes onto the pampas of Argentina, a potential wind farm site Zond has since tried to cultivate.

While the early days of the company featured nights where everyone slept on the floor of the same apartment, Zond would soon rise to the point where its executive salaries would be considered lavish. Still, the firm was reportedly $30 million in debt by 1985. Dehlsen's salary approached the half-million dollar mark, and the Tehachapi wind farms began to creep closer and closer to the power production totals at the Altamont Pass.

◆

Hamrin's day job was energy policy consulting, and her cleints included the CEC and DOE. Since she was gifted in the arena of public presentations and had a way with numbers, she began to represent Pacific Wind & Solar before state regulators, trying to draft terms of their own power sales agreements.

Negotiating these power purchase contracts had become a major headache. With endless parades of lawyers well-versed in delay tactics, only U.S. Windpower and then Zond had been able to get utilities to sign on the dotted line. "I went before the CPUC and showed them a stack of papers that probably stood three feet high and told them that was the amount of documents we had to file over a one and half year period for a project that amounted to a half a megawatt," said Hamrin. Out of sheer frustration, Hamrin and others in Tehachapi put together a new organization, the Independent Energy Producers, in 1982. The group drafted model contracts for wind and other renewable resources, based in part on the work already done by U.S. Windpower and PG&E. The group had an impact. The contracts Hamrin developed were later adopted by CPUC president Bryson. They were institutionalized as the standard offers (SO 4), eventually available to all windfarmers. Even with the utility power purchase contracts in place, Tim Rosenfeld soon soured on the wind farming idea. "I didn't see any turbines that could be trusted, and I got tired of dealing with bankers, lawyers, and engineers." In 1982, $2 million was raised from Hollywood celebrities and others, even though the track record of the ESI turbines was far from impressive. But Rosenfeld and his partners just could not bring themselves to put another wind project together with equipment that wasn't commercially ready.

Thus of the original Pacific Wind & Solar principals, only Hamrin, the company's camp cook, would continue working in the energy business, organizing the wind producers and other small, independent renewable power producers to get better treatment before state and federal regulators. These small operators were more common than either USW or Zond. They needed all the help they could get.

Chapter 14

Sirens and Songs of San Gorgonio

For greed all nature is too little.
—Seneca

The sentries to the San Gorgonio Pass, California's best-known terrain when it comes to extreme winds, are Mount San Gorgonio (11,452 feet) to the north and Mount San Jacinto (10,815 feet) to the south. When they can be clearly seen—which is not often, because of smog—these magnificent peaks mark the boundary between the urban sprawl of the Inland Empire to the west and serene deserts, including Joshua Tree National Park, to the east. For 200 miles, searing winds groan as they flow, unfettered, from sea to desert, terminating at an elevation of only 850 feet in the Coachella Valley.

Cool, moist air originating in the Pacific races to get to the other end of this geological wind tunnel nearly every morning. By the time this air mass reaches the desert, it has become the infamous hot, dry Santa Ana winds that have been known to punish all in their path with an unending flotilla of heat bombs. The only urban outpost of note here in the desert is Palm Springs, which sits in a rare still spot that has been transformed into a playground for folks who look a lot like Bob Hope—folks with white, gray, or no hair who like to play golf and go shopping at places where a pair of shoes can cost more than a TV.

The San Gorgonio Pass features a network of trails first used by the Wanakis Cahuila tribes. Basket weavers like the Kawa iisu tribe in the Tehachapi region, the Cahuila had far more elaborate and more permanent living structures. For example, the ceremonial leader of each clan had what is now called a "fiesta house"; these were considered sacred spaces in which to hold rituals and spiritual dances. Unlike tribes in the Altamont Pass, whose ceremonies often centered on eagles, the Cahuilas considered the largest animals they hunted to be their sacred relatives: grizzly bear, mountain lion, and coyote. Though the Whitewater Canyon became the home village of the Cahuila as their numbers declined from disease, floods have washed away all remnants of their presence on land that now hosts wind farms.

With the help of the Cahuila, William K. Bradshaw, in the 1850s, became the first white settler to discover the windy sand-blasted trails leading through the San Gorgonio Pass. He was in search of the most direct route between Los Angeles and the Colorado River, where gold fever was still in full pitch. All that had passed through this trail before this time were a few expeditions of Spanish explorers, the Mexican Army, and American survey parties. The trail became known as Bradshaw's, and ultimately hooked up with other trails that extended all the way back east to the Atlantic. Bradshaw was credited with opening up Arizona to the rest of the United States. For the next 20 years, the pass was the single link for mail and passenger travel between southern California and the rest of the United States. Later, the trail would also be used for transporting cattle and freight between southern California ports and Arizona stores.

James Marshall Gilman opened the first general store in the Pass in 1871. He later expanded his land holdings into what became the farthest outlying cattle ranch of the San Gabriel Mission. Gilman was an irrigation pioneer and soon had the most expansive ranch in the pass, producing oranges, figs, raisins, prunes, apricots, and olives. In 1909, during the peak of the ranch's fruit production, a Native American laborer named Willie Boy killed his potential father-in-law and in a fit of whiskey-soaked lust kidnapped his intended bride—and then killed her too. His subsequent flight into the desert was popularized in the 1969 film, "Tell Them Willie Boy Is Here," starring Robert Redford. The film depicted what came to be the Wild West's last famous manhunt. James Gilman died in 1916, leaving behind the legacy of the Gilman Ranch, which is now a historical landmark. The arrival of trains in 1890 had diminished the value of the San Gorgonio Pass for the transport of people and freight, but the howling winds would soon attract the attention of California's most ambitious wind power pioneer in the early twentieth century.

Dew R. Oliver, a Texan with a taste for women and the bottle, arrived in California in 1925, sporting a great walrus mustache and a ten-gallon, cream-colored Stetson. An enterprising sort, he immediately started a real estate business in Seal Beach. When it failed to pan out, he began searching for new and hopefully more profitable ventures. While traveling through the rural portions of the Midwest on a business trip, he became intrigued by the small windmills that populated many of the region's small farms. He began to investigate how these unfamiliar contraptions worked. Some ingenious farmers hooked theirs up to automobile generators to charge storage batteries for operating radios. After witnessing these attempts to harness the power of the wind, Oliver began to dream of ways to build bigger and better wind machines to produce cheap electricity. The fuel was free, after all, and Oliver had no money.

When he returned to California in 1926, Oliver discovered a man who was producing small quantities of electricity from a device that harnessed the energy generated by the ebb and flow of tides. This enterprising fellow was actually selling the power he was producing to concessions along a nearby pier. Oliver became inspired and decided to learn more about the electricity business. He also realized that in order to generate consistent electricity, he would need winds far more steady and powerful than those used by the windmills back in the Midwest.

Oliver began to scour southern California, testing wind pressures with his crude instruments, but couldn't find anywhere that measured up to his expectations. It happened that one afternoon he was traveling in the San Gorgonio Pass on another matter altogether—to explore possibilities for real estate. As he stumbled outside to get his bearings, a wind straight out of the west blew his hat off and shredded his road map. He decided to break his journey and spent hours pacing back and forth testing the direction and force of the San Gorgonio winds. His sharp eye noticed how rock walls along the spur ridges of the San Jacinto Mountains formed narrow fingers that reached down to the tiny cities of Cabazon and Whitewater, just north and to the west of Palm Springs. He could tell that these finger ridges were the result of some serious sand blasting that must have been done, he thought, by strong, persistent winds. As far as he could tell, this was the windiest spot for miles around.

Oliver leased the site from a local rancher. After leveling it off, he poured large concentric cement rings upon the ground, upon which light iron tracks typically used in mine tunnels were mounted. Oliver, who had no engineering training, had a vision of creating a gargantuan tube that could channel the invisible wind where he wanted to place a wind turbine. The tube would be mounted on a wheeled platform that could revolve on the tracks, allowing Oliver to aim the tube into the wind, and continue to make adjustments as the wind might deviate throughout the day. In essence, he was further focusing wind that itself was already traveling within a tunnel—the San Gorgonio Pass. The tube he constructed was 75 feet long, a funnel facing into the wind. On the other end, aluminum propellers belted to a generator harvested electricity from mechanically enriched wind fuel.

The first generator Oliver used was one he bought for peanuts at an old Seal Beach roller coaster. The wind was so intense that the propellers spun out of control and the generator immediately burst into flames. Not one to give up easily on what he still believed was a prime business opportunity, he then borrowed a more powerful generator from the Pacific Electric Street Car Company of Los Angeles. Oliver and his cronies held their breath after hooking this massive generator up. The story goes that the new device worked so well that it took Oliver and all of his crew to hold down the generator as it began converting the wild Santa Ana winds into electricity.

The Oliver Electric Power Corporation was quickly incorporated in Nevada in 1926. Oliver borrowed the sum of $12.5 million to startup his business. He began selling $50 shares in what became known as "Oliver's Wind Machine." Never one to shy away from big promises, Oliver had the perfect hook to sell his electricity: The wind was free, how could anybody compete? "Something for nothing" was the kind of sales pitch that resonated in the surrounding desert filled with desperadoes and dreamers. Apparently unaware that the utility monopolies had a lock on virtually all nearby electricity consumers, Oliver announced that the first customer to buy his wind-generated electricity would be the city of Palm Springs. He was a tad late. The Southern Sierra Power Company had captured the Palm Springs account in 1923. Oliver also boasted of other nearby communities that he intended to sell to, but records show that all of his purported customers were already served by other monopoly utilities.

Oliver began running into technical difficulties, too. The wind of San Gorgonio, though remarkably consistent in summer, still fluctuated as temperatures rose and fell with the seasons and from day to day. Oliver's electricity was therefore of varying quality. The individual customers he had signed up for electric service began complaining about the uneven performance of their appliances and lights. A bank of storage batteries could have solved the problem. They never materialized. And while stock sales implied a flourishing business, somehow the whole venture began to head south. Nobody ever seemed to get the whole story from Oliver. But on July 13, 1929, he was charged in a Riverside County court with selling stocks unlawfully and was found guilty and sent to jail, but only for three months (with three years' probation). The sentencing judge had a soft spot for Oliver and let him off easy, having been persuaded to grant the lenient sentence by the testimony of credible engineers that his wind machine was built on sound principles. All it needed to be a commercial success, they reported, was competent management—something that has always been a hard thing to find in the wind farming business.

Oliver was never seen or heard from again in southern California—or anywhere else. Among those investors he fleeced was a widow who lost $6,500.

◆

San Gorgonio Pass is California's best, and one of the nation's premier, wind resource areas. Yet wind farms were installed at a much slower pace here than in either the Altamont or Tehachapi wind regimes. By the end of 1982, only eight small wind turbines were in the ground. By the end of 1984, the San Gorgonio Pass had an even smaller fraction of the on-line capacity of the Altamont or

Tehachapi. Wind farmers didn't stuff the pass with turbines until 1985, the very last year of the wind tax shelter gold rush.

One reason for the slow growth was Southern California Edison (SCE), arguably the most sophisticated, politically adroit, and manipulative utility monopoly in the nation. Although SCE was a pain in the butt to wind entrepreneurs like Zond, it once had a reputation as a major promoter of renewable energy sources. In late 1980, for example, the largest wind turbine ever constructed in California, the 190-foot-tall three-bladed, upwind Bendix Wind Turbine Generator (about the same size as the Smith-Putnam) began generating electricity at SCE's wind energy test center just northwest of Palm Springs. Like Oliver's machine, the turbine, which looked just like a prop-plane engine, sat on a circular track enabling it to be pointed into the wind. Nevertheless, when it comes to independent wind farm projects, no utility has gone to greater lengths to kill them.

In the early 1980s, SCE put only unattractive, "take-it-or-leave-it" power purchase contracts on the table for small wind farmers. Developers had to negotiate a contract with a monopoly utility that one developer described as having "all the marbles." The utility was the only possible buyer (a monopsony): Wind farmers could not sell their power anywhere else, and therefore lacked any negotiating leverage until folks like Jan Hamrin and the Independent Energy Producers came along.

The citizens of Palm Springs themselves were another reason for the slow growth of wind farming in the San Gorgonio Pass. If the Altamont Pass was all about Tesla, CIA insider info, and golden eagles, and the Tehachapi was about that special California entrepreneurial drive and the need to become politically organized at state and federal levels, then the San Gorgonio Pass was about the quirks of local politics and the greed of the final days of the great California wind rush.

In the mid-'80s, Palm Springs made national headlines because of its staunch opposition to wind farming. The City of Palm Springs initially had no direct regulatory authority over wind farms. This infuriated many local residents, who were vocal wind farm opponents. Homeowners living in the long shadow of the wind farms throughout the Pass complained that these strange structures interfered with television reception, an impact of major proportions in a retirement community where wheel chairs outnumber skateboards and it's often too damned hot to go outside. Palm Springs residents, many of them Barry Goldwater Republicans, were incensed that they would have so little say over economic development occurring right in their own backyards. The municipal government itself had a few complaints. "We're working with a developer to put in a six-story resort hotel complex with 200 to 300 condos and a 36-hole golf course here," said Palm Springs City Planner Douglas Evans, a paunchy, fuzzy-bearded man. "Ask yourself," he said, pointing at a wind farm across a freeway, "would you like to look out the window of your condo and see that?"

Projects did nevertheless seem to come out of nowhere. Residents living near the old "Oliver Wind Machine" site near Whitewater were particularly incensed when 30 wind turbines popped up with no warning. "When they start turning in concert you hear a swish and a hum like a large generating plant, which is what they are," said Otis Bainbridge, who built a handsome home here out of scrap wood, glass, and metal in 1978 in an effort to escape the clamor of city life. "Most of us are quite disappointed, to put it mildly." Added a neighbor, who settled in this canyon over 50 years ago: "I call them tin skeletons. They are not natural—that's the whole thing."

A federal agency, the Bureau of Land Management (BLM), issued many of the wind farm permits, since many of the sites were on federal lands. Palm Springs, like Riverside County and other local government officials, could submit comments to BLM, but they served only advisory roles. Exacerbating the tensions between local and federal arms of government was the wind industry itself, whose scheming element was on its worst behavior in San Gorgonio. A wind farming site granted permits for 88 wind turbines ended up with 461; another site with a permit for 15 was crammed with 291 wind turbines. Numbers like these gave angry citizens the impression that BLM was in bed with wind farmers. The frenzied finale of the great California wind rush, the last great American tax shelter, resulted in installations that were clearly pursued for the sake of tax breaks, not power production.

San Gorgonio's development looks and feels like an industrial park. At least, that was my impression that first time I drove through, the bright desert sun reflecting off row after row of spinning metal blades. Instead of being planted along scenic ridgelines, they have in many instances been crammed into the flat desert floor. "This entry once was spectacularly beautiful, a lovely dramatic valley surrounded by tall, majestic mountains," said Palm Springs council member Bill Foster. "Now, however, there is a kaleidoscope of wind turbines that has overwhelmed the landscape and has been as damaging to Palm Springs visually as strip mining has been to towns and villages in Kentucky and West Virginia." From the wind industry's perspective, it seemed like Palm Springs was in bed with SCE. Fred Noble, president of Wintec Ltd., responded to charges of environmental insensitivity by pointing out that the wind farming industry ranked among the top five employers in the western end of the Coachella Valley, with an annual payroll of $6 million.

Two projects by International Dynergy, Inc., of Palm Springs—at Cabazon, one of the most conspicuous sites, and at Maeva—became lightning rods of controversy. The company sold 300 turbines to limited partnerships at Cabazon but erected only 150, and half never worked. Maeva fared no better, generating only 12 percent of its projected output. Bank of America pulled out of the Maeva operation with $1.5 million in unsettled debts, and the IRS paid a visit. True or not, the folklore is that before the visit, employees were seen attaching helicopter blades to machines that had not been operating for months. Worse still was Cabazon's location right alongside Interstate 10, where its heaps of debris were visible to tourists and large numbers of passers by. These two projects helped fuel the public impression that the industry was made up of "tax farms," not "wind farms."

Gary Dodak, a SeaWest employee who began working in San Gorgonio in 1985, has his fair share of wind farming tall tales that locals might not find too amusing. One of his favorite stories involved the Cabazon site. Dodak was out with his SeaWest field crew during a safety training session. Adjacent to SeaWest's property was the Dynergy site, where a Japanese 250 kW wind turbine was being erected. It differed from most wind turbines in that the nacelle and rotor sat on top of a tall, skinny tower that was held in place with guy wires extending from the top of the tower down to the ground. As Dodak's crew watched a crane lift the turbine off the ground, he told them to count the number of safety violations. Apparently no one on the Dynergy crew bothered to bolt down the guy wires near the crane because when the turbine and its skinny tower were pivoted up from the horizontal to the vertical position and releaded, they continued unabated on a downward trajectory, their momentum carrying them straight toward the brand new crane. "You could see the whites of the crane operator's eyes. That's what I'll always remember—his bulging eyes," said Dodak. Luckily, he wasn't hurt when the turbine smashed into the crane.

Maetikhnic, a local southern California outfit setting up shop in San Gorgonio, is another of Dodak's favorites. The company put up five or six wind turbines next to another SeaWest site that were, like the Dynergy Japanese turbines, two to three times the average machine size. One small problem: There wasn't enough room inside the tubular tower to climb to the top to maintain moving parts. So ladders were attached to the outside. Dodak then watched intently as high winds hit. The blades, designed with full-span pitch, rotated or feathered down to reduce wind loads on the machine. Just as a bird tucks in its wings when it wants to reduce wind friction in order to dive bomb, full-span blades can completely collapse in order to avoid the kind of extreme friction that can tear a blade apart. One by one as the feathering blades hit ladder steps they snapped off, sparks flying in the air and people running for cover.

Later, Dodak witnessed a nearly fatal episode at the same site. A teenager was left in charge of the site as the rest of the small crew went home for the day. "The winds started gusting and they had a runaway machine. Instead of going 35 revolutions per minute, it was probably going about 100 revolutions per minute. You could see the kid's little head popping in and out of the bottom of the tower.

Then the first blade snapped off and kind of just froze in the air. Another shot about 300 feet straight up into the air. So the kid started running, but he was heading downwind. I tried to yell to him to turn around, but I don't think he could hear me. I swear it looked like he was trying to catch the blade, which traveled at least 300 yards, but landed within ten feet of the teenager. He just kept running and never came back."

BLM, a federal agency that has never won any popularity contests in the West, approved ten times the amount of wind farm development originally expected. Frustrated by the BLM, Palm Springs took matters into its own hands and filed suit in 1985 against seven wind power companies. A truce in the form of an out-of-court settlement was finally called between Palm Springs and the wind farmers. It required wind turbines to be removed from locations considered inappropriate and prevented highly visible, scenic ridgelines from becoming cluttered with future wind farms. Buffer zones between turbines and roads, with setbacks ranging from a third to two-thirds of a mile, were also to be enforced. Yet the existing visual decay, which included growing numbers of machines beset by mechanical bugaboos, continued to wear down the patience of Palm Springs residents.

Sonny Bono, former pop singer and new mayor of Palm Springs, seized upon what seemed to be a hot button issue among his local constituents in the summer of 1989. He traveled to Washington, DC, to lobby against these irresponsible wind farmers. Bono called upon Congress, or somebody in the federal government, to grant Palm Springs a greater voice in stopping nearby wind power development. Wind farms were a major threat to the local tourism industry, he charged. Bono came away disappointed, learning from the Department of Interior that the law was the law: BLM, not Palm Springs, had jurisdiction over wind farms on its lands.

Just over a year later, Bono had changed his approach. Palm Springs was in the thick of trying to balance its budget. Bono threatened to annex wind farming properties as a way to increase its tax base. This proposal had the wind industry up in arms and further heightened the animosity between the city and the wind industry. A coalition of wind farmers then began negotiating with Bono behind closed doors. Given the lack of industrial activity in this retirement community, and the growing role wind farmers were playing in local economic development, Bono suddenly declared the war between the wind farmers and the city over. At a hastily called news conference, Bono conceded he "switched horses in midstream," adding that "what started out as a problem turned into what I think can be a tremendous asset for the city." He went on to praise the wind farmers, whom he had just lambasted in Washington, DC, for their major contributions to the local economy. "This is a multi-million dollar business at the doorstep of the city," Bono acknowledged. He pointed out that by including the majority of the 3,500 wind turbines within the city limits, Palm Springs would net $1.6 million in annual property tax revenues. On top of that was $1.8 million in sales tax revenue every decade.

Palm Springs called off its fight against wind farms in exchange for their compliance with local guidelines in siting new machines. "This is a new beginning," said Noble, the local voice of the wind farming industry. "The city wanted to annex our land. Given their attitude toward us, that would have precipitated a major battle," he added. Noble went on to say that ending the legal battle between Palm Springs and the wind industry was in everybody's interest. "Despite our best efforts to fight at the county, state and federal level," summed up Bono, "wind turbines are not going to go away. My feeling is the best way to go is not to try to beat them, but to join them."

◆

SeaWest survived the extremely tough wind business because of its ability to juggle several financial balls at once. "The IRS went over all our projects and never disallowed one," said Chuck Davenport, a SeaWest principal, with a tinge of pride, highlighting his own measure of a wind company's success. Unlike USW, SeaWest never designed or produced any of its own turbines. Vertical integration seemed to be a recipe for disaster, given the long list of technology failures, said Davenport, the CPA whose first experience with a wind turbine manufacturer was the tragic Terry Mehrkam story. Other U.S. companies "fell in love with their product and never perfected their wind turbines. SeaWest, thank God, never had enough money to enter the manufacturing business."

Though SeaWest's first wind farm was set up in the Altamont Pass in 1982, Davenport's company (called California Wind Energy Systems until Davenport changed the name to SeaWest in 1984) didn't really take off until 1983 when it hooked up with Kidder Peabody to finance wind farms in the San Gorgonio Pass. Whereas SeaWest used American-made Enertech turbines in the early years, it later relied almost exclusively on Danish, Japanese, and other foreign models.

"We created the Danish market for wind power in San Gorgonio in 1985," Davenport said in a rare boast, focusing, as he always does, on the financing side of business. New tax laws in Denmark created opportunities for Danish investors to gain tax benefits by investing in new California wind farms, and Davenport was the first to tap into this new source of investment for SeaWest projects. (Interestingly, Dehlsen told me he had been honored in Denmark for his efforts to bring Denmark into the U.S. market for wind, on the basis of his deployment of large numbers of Danish turbines.) Davenport's perspective on bringing Danes into the U.S. market focused on new investment opportunities for Danes themselves in California. "The last project with Danish turbines and

Danish investors was in 1988, when I replaced the old Mehrkam wind turbines down by San Diego," he recollected.

Davenport, like Dodak, observed that most American machines couldn't cut the mustard in the San Gorgonio Pass. They were too lightweight for the howling winds that would regularly rip through. SeaWest made a business out of retrofitting older U.S. and Danish machines. "Salvaging old technology is a valid base of business," said Dodak. "We've performed upgrades we thought would keep the turbines running for three to five years, but we got seven to eight years instead. But we are starting to replace a number of the older turbines at our projects with new stuff. There is just no such thing as a 20-year fix," he said. Dodak, too, thinks that SeaWest's approach to the wind power business makes an awful lot of sense. Two or three times in its history, SeaWest looked at forming joint venture turbine manufacturing partnerships, but each time it backed away, he said. "All wind power sites are very different. What might be optimal in the Altamont Pass may not be optimal near Palm Springs. Once you build a wind turbine, your project development team then has to use that turbine for all of your projects, even though the turbine may not be well-suited to a particular project. Some manufacturers offer warranties, which might be important to some investors. Others may offer hi-tech or give you more energy value per buck at high-wind sites."

Davenport, good CPA that he is, listed a series of financial transactions, sell-backs, technology transfer agreements, and other clever money maneuverings that have kept SeaWest afloat while other peers sank. "We've had our slumps. The key was we were never a wind turbine manufacturer."

Chapter 15

Shelter from the Storm

> Technology can accomplish anything in 50 years; in five years, it can accomplish nothing.
>
> —Edward Teller

California's three primary wind farming regions—the Altamont, Tehachapi, and San Gorgonio—served as a huge wind technology laboratory. Some of the biggest skeletons continue to litter the landscape, often cannibalized by the rugged forces of the wind and the equally devastating impact of unforgiving financial markets.

The severity of the problems with USW's first-generation turbines was documented in a PG&E report, and is emblematic of the industry in those first years: "As of June 30, 1984, 179 of the 557 early-version model of the Model 56-50's had failed in the field and an additional 352 had been shut down because continued sustained operation entailed significant risk of severe damage to these windmills." The report was authored by Don Smith, described by one wind industry veteran as a "pony-tailed, whiskey drinkin' neo-Marxist." The soft-spoken Smith meticulously analyzed the performance of USW machines, as well as all other wind turbines that operated in the Altamont Pass.

While Smith's data had documented the toll the Altamont Pass's winds had taken on USW's first-generation machines, his figures also showed that the 529 of its current and improved 100 kW models appeared to be the Altamont Pass's top performers in 1985. Nonetheless, he pointed out that USW had installed so many turbines so rapidly that exact figures were unavailable. Turbines with variable pitch, the vast majority of which were USW machines, met 73 percent of their projected 1985 production. Fixed-pitch units, such as Fayette's, produced only 43 percent. (It was later discovered that USW's reporting techniques may have skewed Smith's statistics, since their one-of-a-kind strategy was to guarantee a certain amount of output to their limited partners; the company would then install as many turbines as necessary to meet these production commitments.) But Danish turbines, such as those used by Zond, and other European turbines were on the whole much better performers than U.S. machines: They averaged 69 percent of projections; U.S. turbines generated only a little more than half of what was expected of them.

Even the stout Danish wind turbines, designed to withstand a hurricane, lost in their battle with California's erratic and unpredictably turbulent winds. The blades of these turbines, while lasting far longer than their American counterparts, were too stressed by the brutality of the California wind regime and started failing much sooner than expected. Indeed, all of the blades on these machines needed to be replaced in a comprehensive retrofit. Each of California's major wind farming regions had its own peculiar performance and fatigue challenges in addition to the sheer power and volatility of the wind. In the Altamont Pass it was bugs, whose build-up on blades could reduce energy production by 15 to 50 percent. In Tehachapi it was snow and ice that could freeze parts, including rotors, which then cut into performance during winter. In San Gorgonio, wind turbines were literally sandblasted, which wreaked havoc with gears and scarred blades, again reducing power production.

Unlike most power-generation technologies, the experimental failures of the wind program, both California's and the nation's, are very visible to the public. One can clearly see whether they are working or not. They stand out because they must be placed out in the open, in conspicuous spots where most people are not accustomed to seeing power plants. Since California's wind season is the middle third of any year, there is a good chance the turbines will not be spinning away when the average citizen drives by. There are times of the year when the turbines operate only at night, when nobody is around to witness the transfer of kinetic energy into calibrated units of electric power. The turbines sometimes form stark silhouettes on the ridgelines. All many see are a bunch of odd, sometimes rusting contraptions. Many visitors from afar, nonetheless, have come to view the turbines as an extension of California's mystique, a museum of sorts that depicts the evolution of a technology nurtured by that special Golden State frame of mind.

According to a University of California study, the notorious state tax credits that Cashman convinced former California Governor Jerry Brown to offer to the wind industry raised $528 million in investment dollars. Federal tax credits added another $880 million of private investment. Yet, many electricity industry analysts and insiders think the $200 million annual losses in federal and state taxes that occurred primarily over a five-year period in the early to mid-1980s—totaling about $1 billion—was money well spent on developing wind power. It is but a blip compared to ongoing tax losses to other, far less benign sources of power. "Wind energy is a tax bargain," writes Thomas Starrs in an award-winning article in the *Ecological Law Quarterly.* He estimated that at least a billion dollars in annual tax subsidies had been afforded the fossil fuel and nuclear industries over a period well beyond 30 years. In 1984, for example, the nuclear industry was the recipient of $15 billion in subsidies, whereas all of the renewable energy industries received $1.7 billion. The U.S. Congressional Research Bureau reports that 65 percent of the nation's R&D dollars spent between 1948 and 1990—$32 billion—was earmarked for nuclear power. This compares with 11 percent, or $5.6 billion, for all renewable energy sources combined.

By the end of 1981, 144 wind turbines were installed, representing 7 MW and generating only enough electricity to meet the annual needs of two households. But, because of the tax credits, wind turbine technology achieved the maturity in the next five years that typically takes 15 to 30 years in secluded government labs. Utility research and development scientists and engineers at the Electric Power Research Institute (EPRI) have been bullish on wind power for quite some time. Ed DeMeo, former EPRI manager of renewable energy programs, notes that the use of tax credits was "far more effective than the federal wind R&D program. Though not perfect, the credits helped improve the technology bit by bit."

In a 1987 report entitled *The Wind Farms of the Altamont Pass,* Don Smith calculated that the Altamont farms alone produced half of the world's wind-generated electricity in 1985. The Altamont wind turbines would produce about 550 million kWh in 1986, Smith said, adding that this amount of electricity generation was the "equivalent to the residential electrical needs of a city of about 250,000 population, and about 1 percent of the electrical energy used in northern California."

Not everyone was impressed. Lou Divone of DOE complained that the state and federal tax credits generated "obscene profits," since the size of tax write-off was based "on how much you spent—not how much energy was produced." "The way in which the tax credits were implemented invited abuse," added Andy Trenka, currently a professor promoting sustainable business development at Colorado State University. Trenka previously ran the DOE Rocky Flats testing program that convinced Ty Cashman that California had to do something on its own with state tax credits. "All of the benefits went to the investors when there should have been a sharing of mutual benefits among all partners, including the electric utilities who were required to purchase the wind power," said Trenka. A modern industry was born in California largely due to the investments generated by the tax credits. Even Cashman acknowledged there were problems. But he blames many of the drawbacks of the tax credits on President Reagan's deregulation of the savings and loan industry. "I thought stodgy bankers would put a brake on the program," Cashman stated. Instead, S&Ls,

hungry for strange new places to put money, accelerated the use of the tax credits. "I always knew it was going to be messy," says Cashman, "but money became available for poorly engineered machines. Badly conceived wind farm projects were also rewarded."

Though Cashman's goal was to give wind technology a big push forward, Trenka claims that part of the problem of accumulating wind turbine performance data at Rocky Flats was that the effort to commercialize wind power technology "went too far too fast in the early years. The wind turbine manufacturers kept saying 'We're commercially ready. All we need is a little bit of a push to achieve volume production.' But in reality, the wind turbines were not yet ready. We couldn't keep the turbines running long enough to get good performance data," Trenka argued.

Looking back, Cashman acknowledged that "the Danish saved my butt—they were the white horses. The old path of technology was heavy gears and heavy weights and that was the Danish approach." Since the Danish subsidies for wind power were linked to weight, designers there tended to err on the side of making machines heavier, not lighter. The new path, at least from Cashman's view, was influenced by the Zen notion of flexibility. "The image of bamboo comes to mind. With a big wind, the large oak tree might crack, but bamboo, because of its lightness, could bend and touch the ground," said Cashman. Many of the U.S. turbine designers had worked at Bell Helicopters. "These designers had the notion embedded in their brains that lightness was a good thing, too. After all, helicopters have to get off the ground. With wind power, however, you don't have to get off the ground."

In retrospect, Cashman thinks the California tax credits were "much too generous—ridiculously so." Still, he has no regrets about the policies he helped successfully carry out in California. "We were stuck and I threw a stick of dynamite to break open a vicious cycle that was killing us. Nobody wanted to invest in new energy technologies. What we did was make it so seductive that they would invest—even if the wind turbines didn't work. Without our program in California, the cultural knowledge would have been that wind power doesn't work," said Cashman. Though the backlash against wind power grew in part from the perception that wind projects were built more for tax benefits than for energy production, Cashman's current concerns about global climate change, and the present boom in wind energy as a response to this growing crisis, bolster his view that the investment tax credit positives outweighed the negatives when one looks at the long-term big picture.

While the investor tax credits are often a favorite topic for a toast by wind pioneers reminiscing about the good old days at the annual meetings of the American Wind Energy Association, Divone's program at DOE is a common object of scorn. Aerospace companies like Boeing, Lockheed, Westinghouse, and General Electric were given fat DOE contracts to design a series of Putnam-scale machines while medium-sized wind turbines (at the time, the product most appropriate for California wind farms) were dropped from the testing program. In contrast, the Danish approach focused more on helping small private companies, many of which used to design and build farm equipment, meet the needs of the rural cooperatives and other farmers that wanted to buy wind power with much smaller models before scaling up to turbines two and almost three times larger than the Smith-Putnam machine.

Which approach worked better? Consider these facts. Between 1974 and 1990 U.S. taxpayers invested $450 million in research and development funding in an effort to develop a superior gigantic turbine that would appeal to utility monopolies. According to Woody Stoddard, the United States knew no more about wind power after that half-billion was gone than the Nazis did under Hutter some 40 years earlier. Danish taxpayers, in contrast, invested about $52 million, most of which helped develop much smaller-scale machines that could be used to supply power to rural farms and villages. Not one commercially viable machine emerged from the U.S. federal government's forays into developing a wind turbine based on sophisticated aerospace engineering science and the pursuit of greater efficiencies. In Denmark, it was the "uneducated" metal and machine shop workers that came up with the best turbine designs, which were smaller but extremely sturdy, featured three blades, and faced into the wind.

Though the United States prides itself on its faith that the free market separates winners from losers, most of the early American wind turbine development contracts were handed to institutions that were sheltered from the discipline of real market forces. These were huge companies that built bombers and nuclear reactors. They didn't have to worry about surviving in a real market setting since much of their past work had been heavily subsidized and had only one customer—the military or a utility monopoly.

In a paper he prepared for Stanford University's Department of Economics, Peter Karnoe of the Copenhagen Business School argues that the U.S. federal wind program was doomed by "typical" American weaknesses. Karnoe is right. The Danes were "copycats," since nearly all manufacturers relied on the same three-bladed upwind machine. So there was no competition on fundamental design issues. Instead there was cooperation in securing common parts, and engineers were devoted to incrementally improving power production by focusing on refinement of details. The United States, in contrast, had designers embracing a much greater diversity of turbine approaches, each trying to make the ultimate one-of-a-kind breakthrough in some radical new way to generate electricity from the wind. There were vertical-axis machines, machines with two or three blades, mostly facing downwind, but some upwind. The greatest efforts by Americans were in finding ways to reduce weight, often experimenting with blade materials and shapes and sizes.

To Karnoe, the U.S. industry looks like a bunch of suspicious and secretive entrepreneurs who never shared any lessons learned from the field and therefore trailed the Danes when it came to steadily improving performance from wind turbines being installed in California.

The earmarking of Uncle Sam's funds to big firms to enable them to develop big wind turbines should come as no surprise. The Energy Research and Development Administration (later to become the Department of Energy) traced its institutional culture back to the Atomic Energy Commission, which was long ridiculed for its relentless promotion of nuclear power. These federal bureaucrats shared the management culture of utility engineers, white males who had become mesmerized by the notion that "big is beautiful," in the words of Carl Weinberg, a former PG&E executive who managed the utility's research and development efforts in wind, solar, and other clean, futuristic power generation technologies.

Jamie Chapman, a private consultant once employed by USW to make its first-generation wind turbine work in the field, defends both the federal program and the private sector tax credits. "They were two parallel paths and, at the time, were both reasonable things to do. The federal government saw utilities as the prime customers and the thought was that they would be most interested in very large machines. Bigger is better in their eyes. The problem was that once there were problems, it cost way too much to retrofit." Chapman argues that though none of the big machines supported by the federal government became commercial designs, the research did advance the state of knowledge. "The tax credits, both federal and state, were essential. They put wind power on the map and attracted the private capital that gave companies the cash flow to design, manufacture, and install machines—and then fix the problems. Because of the tax credits, we now have modern wind turbines with no structural failures, that can operate in high-velocity wind regimes and are quite cost effective."

Others are less charitable. Among them is another Chapman, William Chapman, who is all that is left of WindMaster, which set up shop in the Altamont. Not one to mince words, he told me that the people who designed the tax credits "had their heads screwed on backwards. It was all based on how much concrete was in the ground, not how many kilowatt hours were generated. It was the investors who got all of the credits and made most of their money in the first five years. We now have, at the taxpayers' expense, one of the largest displays of free form art in the world." While he acknowledged that the wind industry, like any new technology, needed to go through a growing stage, he lamented "there is no support today to go to the next plateau." This Chapman directs his sharpest barbs at state regulators. "When the utility monopolies hiccup, the regulators come running with a Kleenex. When the wind industry needs help to keep from dying, they shrug their shoulders and say, 'It didn't work, did it?'" He concluded, "In the 1970s, the projection was that electricity would be costing 13 cents a kWh today. The only entities getting those prices for electricity are utility nuclear reactors. Now what does that tell you?"

◆

Both federal and state credits were repealed in 1986 after revelations of shady deals and unpaid bills. In one case, San Diego area residents, including numerous retired school teachers, invested over $6 million into a local wind farm. Among them were Barbara and Rainmundo Neri, who invested $5,000 in a single-windmill limited partnership with promises of profits. The couple ended up losing $20,000 in back taxes, interest, and penalties.

In another case, a company called Trans Power sold windmills with an appraised market value of $3,000 for $300,000. The result was, among other things, tax losses to federal and state governments amounting to $2 million, according to court records. The company, like many instant firms that popped up during California's wind rush, vanished once its scam was discovered by the authorities.

Congressman Pete Stark of Hayward, in the Bay Area, led the charge to defeat the extension of the tax credits. "These aren't wind farms, they're tax farms," he berated Congress. Before any more dollars went out the door, "let's separate the wolves from the sheep," he told a receptive Congress.

Denis Hayes, who helped set up the first Earth Day and is a long-time advocate for renewable energy, strongly disagreed with Stark's rhetoric and made the best speech on behalf of extending tax credit support for the renewable energy industry:

> Congress claims to care for the environment. But it proposes to pour buckets of money into energy that rapes the landscape, pollutes the air, causes a global greenhouse effect, and produces dangerous isotopes that will still be in the first half-life in the year 150,000 A.D. Meanwhile, it designs policies that will drive benign and sustainable energy sources from the marketplace.
>
> Congress claims to be worried about the budget deficit. Yet a century after Drake sunk the first oil well in Titusville, and three decades after Eisenhower launched Atoms for Peace, the fossil fuel and nuclear industries receive yearly subsidies that could displace one-fourth of the federal deficit. The current tax bill winks at an elephant while it crushes an ant.

Bob Thomas, the Jung enthusiast, still staunchly defends Cashman's tax credits, too. "When all is said and done, the key to technology advancement depends on how much capital you can get absent government financing. To get people to part with their money you have to cut the risk of loss and rely on greed as a motivating force," says Thomas. He argues that it is difficult to motivate an industry from pure altruism. Though there were plenty of mistakes early on, every technology goes through "birthing pains" and a learning curve before maturing. Costs, including scandals, are always a factor when one relies on the mad humans who practice the religion of high finance. The costs are high to start any industry, Thomas reminded me. "The shadow tends to intervene when you set up such programs," Thomas remarked, referring to the dark side of human nature. "Hopefully, the shadow doesn't dominate, and there is plenty of creative and positive light," he said.

Chapter 16

All in the Family

A family is but too often a commonwealth of malignants.
—Alexander Pope

Enertech already had its wind turbine business up and running in New England long before USW installed its first wind farm in New Hampshire, though its focus was on small turbines. In the early 1970s, long before the term "oil embargo" had become part of the country's everyday vocabulary, Vermont native Bob Sherwin taught an energy conservation course at Rutgers University, in New Jersey. Sherwin loved ships, trains, and windmills as a kid and was still fascinated by the engineering innovations integrated into these marvelous inventions. After teaching for seven years he grew tired of the New Jersey commute. He happened to own some land that didn't have electricity. When the energy crisis struck in 1973, Sherwin began to toy with the idea of generating power from the wind. He and a friend who worked for NASA launched Enertech Corp. in New Hampshire in 1975. Solar Wind Co., a small company in Maine that imported wind turbines for domestic sales, was snatched up by Enertech to provide a wind turbine product it could sell. Initially focusing on off-grid agricultural markets, Enertech began receiving orders from all over the world and quickly grew into a $2 million business.

Sherwin stumbled upon an invention that changed the business strategy of the young company. Xerox hired Enertech to create a bike power system for children's museums that would show kids how much human energy it took to generate electricity. In the process of designing the system, Sherwin developed an induction generator, which produced AC current. Enertech then adapted this device for use with wind turbines so they could be easily tied into the utility grid. Enertech began designing its own wind turbines while it continued to sell imported turbines. (The firm's first commercial sale was, ironically enough, to USW's Russell Wolfe.) Unlike USW, Enertech was aided by two Department of Energy contracts that totaled $2 million. At 1.5 and 4 kW, these skinny toothpicks-of-a-turbine were tiny compared to the 50 kW machine that USW would first put in the ground.

Both companies' turbines nevertheless shared typical American design fundamentals: downwind, free-yaw, variable-pitch blades. It was the rotor that was different. Enertech's was fixed pitch and stationary, and USW's was variable pitch, moving in response to the direction of the incoming wind. The Enertech turbines performed reasonably well. They were the state of the art in the 1970s. By 1978, Enertech was putting out 100 wind turbines per month and generated over $200 million in sales, as its machine was the only utility-grade wind turbine on the market.

Sherwin looked like a shaved Santa Claus when I visited him on a November day in Vermont; his most distinctive feature is the twinkle in his eyes. He too recognized the importance of bringing mainstream corporate financial firms and investors into the wind business. On the financial side there was Kidder Peabody, which worked with Sherwin to install more than 750 wind turbines in California. In 1982, the Bendix Corporation extended a $2.5 million line of credit, a large portion of which was invested in retrofits to correct blade problems on Enertech's turbines. But when Bendix was purchased by Allied General, the new owner placed a hold on the financing as it closed down its environmental divisions. That marked the beginning of the end.

Sherwin lost his shirt ("millions" is what he mumbled) and walked away from Enertech in 1984. As in the designs of many other wind turbine manufacturers, the hub plates that connected the blades to the rotor cracked in the insane California wind regimes. But he insists today, as we sit in an office that is a converted barn at dusk on a gorgeous winter day in Vermont, that the failure of Enertech "had nothing to do with the design or the hardware." He claimed the money guys were to blame, despite reports that the passive yaw design in Enertech's larger turbine models concentrated stress at a point on each tower leg that resulted in tower failures. "We had tower problems in one project in San Gorgonio, but that was caused by stuffing too many turbines in too small a space—a decision made by the developer, not us," said Sherwin, defending the engineering and drawing a sharp distinction between the technology and the business ends of wind farm operations.

Enertech was almost rescued by Westinghouse Corporation in 1985. The corporate giant had approved a rescue plan, but came up with the cash one day too late. In the process, Enertech gave up a lease on key wind farm sites in the Altamont Pass to USW. "I know it's serendipity now, but Enertech should have, and could have, survived," said Sherwin wistfully, before a long, great silence as the sun started to set, turning a fiery red orange. He motioned to an old rusting wind turbine located here, at the headquarters of his current company, Atlantic Orient. "That's the original Enertech turbine," he said proudly. "For the hundred or so folks who worked at Enertech, it was the best times of our lives. We were one big family, all dedicated to pushing the frontier."

In the beginning of the wind rush of the late '70s and early '80s, all of the initial domestic choices for wind turbines were lightweight designs geared for off-grid applications or for on-site power generation. Before new laws enabled the independent power industry to sell its electricity to the utility monopolies, these lightweight American machines represented an evolution in design that traced its roots to Marcellus Jacobs. The companies were a family, albeit a feuding one, in the sense that they all faced the same challenge: figuring out ways to make small, lightweight American wind turbines work in the harshest of California's wind regimes.

◆

Enertech was lucky to tap into the funding that Divone provided through DOE's small wind turbine development program. Sherwin was part of the New England "mafia family" that Ty Cashman once derided. Most Midwesterners developing wind turbines, however, relied on their actual families to get their small businesses off the ground. For example, Jay Carter Sr. and Jr. started Carter Wind Systems (CWS) in 1976, in the little West Texas town of Burkburnett, just south of the Oklahoma border. The small company was indeed a family business: Over 75 percent of it was owned by the Carter clan. Carter Sr. put up most of the initial money. Along with Terry Mehrkam's and Enertech's, Carter's machines represented what the domestic wind industry had to offer independent developers hoping to make utility sales at the start of California's wind rush. Randy Tinkerman chose a Carter turbine for his first California installation, not too far from the Altamont Pass. It appeared to him to be the most promising of the bunch, since it looked like it could be scaled up to larger, utility-grade machines.

The Carter designs are everything the typical Danish machine is not. They take the concept of light weight to incredible extremes, weighing one-third as much as the average Danish machine—and costing 30 percent less. A wind farm using these lightweight machines leaves few environmental footprints. The turbines can be raised and lowered simply, since no cranes are needed. Nor do they require heavy equipment, and consequent service roads, that can cause erosion. The greatest selling point of the technology, however, is that its tall towers may be ideally suited to the Great Plains, where 90 percent of the nation's wind resource is found. Wind shear increases with height, and a lightweight turbine can capture the available energy that exists at higher elevations without the prohibitive capital costs of heavier tall towers.

Or so say the advocates of the Carter machine. The Carters themselves aren't very bashful about why they think their designs are keen. "Three-bladed machines are going to go extinct," proclaimed Carter Jr., a balding, gregarious man who radiates a homespun Texas swagger. "They will go extinct, just like the dinosaur went extinct. Not that Danish machines are not good. Not that they are not strong, reliable, and produce electricity. But the bottom line is: The strongest survive. A three-bladed machine cannot compete with a two-bladed machine."

The two flexible fiberglass blades that can shed extreme wind loads much as a palm tree handles a hurricane are the most striking feature of the Carter turbine. In fact, the blades can bend 60 degrees in winds of 160 mph. The blades are hollow, but a spar spinning inside the blade is what actually curves, explained Carter Jr. "Fiberglass can bend. I mean, you can bend a fishing rod completely in a circle if you wanted to. Now if you make that fiberglass into a tube, you can't bend it. That's what we've done with our blades. Each blade is then held at that fixed pitch by a strong electromagnet located inside each blade. And each blade is independent of the other. If something, for heaven's sake—if something should ever happen to one blade and it won't pitch up, the other one can still pitch up and stop the machine."

Unfortunately, CWS never reached the sales figures of Enertech. It is one of those companies whose time line is peppered with near misses, and a few disasters. The low point was when the company's blade shop burned to the ground. The doors were closed on CWS completely at one point, though bankruptcy papers were never filed. Then came the family feud.

Carter Sr. began to work on a new blade design, telling everyone that the original Carter design he helped put together, and was still being promoted by his son, was a "crock." Carter Sr.'s new turbine came to be known as the "Wind Eagle," and the changes he made in blade design were prompted by what he witnessed in California. During one wind storm, "where those mountain winds came spiraling like a tornado at 100 to 110 mph," Carter Sr. watched a blade bend over so far that it hit its own tower and self-destructed. The independent blades, pitching in different directions at the same time, had the poor machine "oscillating back and forth." In his "Wind Eagle," he tied the two blades together and used hydraulics and powerful springs to pitch the blades up. The original design relied on centrifugal force and an electromagnet to do this. With the blades fastened together, the teetering hub no longer worked as an efficient method of gathering energy, so Carter Sr. went with a rigid hub, and as a result the whole machine teeters. So, the son's turbine rocks side to side, the dad's moves up and down.

Despite their brilliance, both Carter wind turbine designs have so far been held hostage by their makers. Both have refused to finance a detailed third-party verification of their performance claims. There is also still some work to do when it comes to standardizing the specs to achieve the quality control required for mass production. Both Carters have been, and probably always will be, backyard tinkerers. They are rebels with a cause—but can't find anyone to put up with their idiosyncrasies and their huge inventor's egos. The dissolution of their partnership only hindered the chances for either one to succeed, since they both continue to bash each other's turbine instead of promoting the overall turbine architecture on which they still see eye-to-eye. Robert Kahn, a former public relations consultant who worked with both Carters, had this to say: "They had the whole world in their hands. It is a tragedy. They had such a cool, revolutionary design. It was the classic tale of the inventors who couldn't let go."

◆

After working at Piper Aircraft for a number of years, Karl Bergey began to teach aeronautics and mechanical engineering at the University of Oklahoma in the late '60s. In 1970, in light of the so-called energy crisis, he received a small grant to perform a feasibility study to help determine what role wind power might play in displacing fossil fuels. Oklahoma's economy was clearly dependent on the oil and natural gas industry, but steady winds came rustling off the plains, and Karl thought there was enough wind to spin turbines to help pump oil out of the many wells that peppered the landscape. Meanwhile, his son Mike was busy studying for his mechanical engineering BA when he found out about a student design contest for wind turbines. Mike decided to get involved. "I was a gofer the first year, in 1975," said Mike, adding that his team's machine that year failed to drum up much enthusiasm. But this didn't deter his interest in wind power. Quite the contrary. His fascination with engineering a new technology such as wind turbines just kept growing. "So the next year, I was in charge of my own design and I went with a vertical-axis machine," said Bergey. Unlike the far more common horizontal-axis machines, a vertical axis is omnidirectional. This was viewed as a major plus since wind from any direction could be easily accommodated and converted into electricity. Because they are installed at ground level, vertical-axis machines are also easier to maintain.

Bergey's design was unique in another way. Instead of the straight blades used by San Rafael-based FloWind, the sole major U.S. firm that had staked its success on a vertical-axis Darrieus design, Bergey's blades cyclically changed pitch in relation to the wind. "I calculated that my 8 kW wind turbine was 59.3 percent efficient, which was the most efficient of any turbine entered into the contest, because I got two bites from the wind," said Bergey. He had designed his blades to articulate in order to capture more energy from the wind. Not only did the younger Bergey win the contest, but he soon discovered another form of flattery. McDonnell Douglas, with funding from DOE, developed a 40 kW "Giromill" vertical-axis wind turbine. "They crawled all over my machine and then came up with something that looked remarkably like what I had designed," said Bergey, a small smile escaping from his broad Norwegian face.

Whereas most of his friends found lucrative jobs in the booming Oklahoma oil business, Mike was convinced that wind power "was going to change the world." He graduated from college in 1977 and immediately he and his dad began collaborating to design a wind turbine they hoped to manufacture and sell to homeowners, farmers, and yes, oil companies. "He was chief and I was Indian," recalled Mike. Karl had pointed out that the mechanisms that changed the pitch of blades in Mike's winning turbine design were too complex and often generated unwanted drag, robbing the turbine of potential power production. Ultimately, the vertical-axis design was dropped in favor of the more traditional horizontal configuration. In spite of theoretical high efficiencies, "all of the vertical-axis machines were notoriously unreliable," Mike recalled. "I have to give my dad full credit," he added. "For three years, he continued to hammer away on the notion of getting rid of as many moving parts as possible. Our first turbine, a 1 kW machine we completed in the summer of 1980, was still fairly complex. But we just continued to simplify." Indeed, Karl liked to describe the proper wind turbine design philosophy by quoting Antoine de St. Exupery: "Perfection is achieved not when there is nothing more to add, but when there is nothing more to take away."

Bergey's design was revolutionary in the 1980s. "Now it is just the norm," he shrugged. Mike added, "There are no brakes, no controls in the nacelle. We used a simple, light generator and created an extremely simple blade-pitching technique. Our Bergey wind turbine was more like the original Jacobs machine than his second design—the one he designed for AC grids. We ended up following Jacobs better than he himself did." In the late '70s and early '80s, wind energy pioneers were at work at the Red Wing Institute in Minnesota; a company called Windworks in Wisconsin spent a lot of time refurbishing old Jacobs machines; and then there were the New Englanders like Sherwin who also were designing small-scale wind turbines that were now being installed in large corporate wind farms. By far and away, though, Jacobs was the best-known wind farmer in the land. "Marcellus could have been the father of the wind power resurgence. But he had the wrong personality. He just couldn't work with others. He wanted everyone else in the industry to take his word—and his word alone—as gospel," said Bergey, shaking his head. "He would call all of us 'whipper-snappers,'" he chuckled. "His whole livelihood was killed by Washington, DC, the Rural Electrification Act. That's what did him in."

The California wind rush fueled the rapid expansion of the wind power industry, but it also shifted the focus away from the engineering of small-scale American turbine designs to the seizing of once-in-a-lifetime business opportunities. The goal was to install as many wind farms as possible, with bigger and bigger machines, in order to take advantage of tax credits that certainly wouldn't be around forever.

Chapter 17

Not in My Backyard

> Patience! Does the wind mill stray in search of the wind?
> —Andy Sklivis

Those companies that survived the end of California's tax credits had no choice but to refine or completely rethink their business strategies. U.S. Windpower chose "Kenetech" as its new moniker in June 1988. According to upper management, the new name implied "a moving force in the development of energy systems and technologies." There was another reason for the name change, however: The company was beginning to seek opportunities in other countries. The old name was seen as a liability in those foreign markets. For the first time in its history, the company was thinking of diverging from the vertically integrated business strategy by which it rose to the top: providing all of the products and services necessary to generate power from the wind. Now it was looking to sell its wind turbines to other developers; to build projects for third parties, including partners in China and Spain; and to use its expertise in a consulting capacity for utilities and other developers.

On paper, USW had looked to be in peak health. Although its payroll shrank in 1987 to 300 employees from 460 the previous year, the firm had a net income of $8 million on $81 million in revenue, up from its $5.3 million profit on $84.6 million in 1986. And its assets showed a major increase—$226 versus $139 million in the previous year.

Then, in 1988, Dale Osborn was hired as president, replacing Gerald Alderson. Alderson had replaced Moore as president back in 1981, but didn't really distinguish himself at the company until after the tax credits ended and the company pulled away from the rest of the pack in terms of sales and revenue.

Alderson's contributions had been critical in moving the company from the status of early pioneer to survivor and then industry dean. Osborn called Alderson "the spiritual leader" of the world's leading wind farming firm, someone who had set a clear-cut course by which "we didn't have to rely on anyone else for our success." USW's vertical integration had served the company well in the early days.

But Osborn was trained in a different environment and had different ideas. He spent a couple of years as vice president of marketing for Public Service of Indiana after spending 13 years at Texas Instruments. His experience in the computer industry in which aggressive cost-cutting was a management mantra defined the course he would set for Kenetech and to which it had now publicly committed. "We have a new vision of how to get things done," Osborn proclaimed, "and it is called 'design-to-cost.' We have set very specific cost goals for each component of our new wind turbine—assuming an Altamont Pass wind resource. I've told the engineers: 'If you can't get there, attack the problem until you get to the target cost.'"

"This is not a fraternity, this is a highly competitive business arena—and we are now the most competitive organization on this globe to deal with wind issues," said Osborn. Acknowledging that he and his colleagues had been accused of being arrogant, Osborn brushed aside such criticism, asserting that the perception merely reflected the fact that the company always did things its own way: "We've been our own player." Since the company did not need to rely on any other manufacturers or service companies in the wind business, Kenetech was seen as being isolated from the greater wind farming community at large. But that was about to change, Osborn promised.

In 1989, the firm finally gained approval of a project in Solano County—the Montezuma Hills wind farms. While other developers had failed to gain approval for projects in Solano County, Kenetech succeeded by choosing a site that was far from human populations and by agreeing to a number of environmental mitigation measures, including providing nesting boxes for birds and using owl decoys to dissuade birds from getting too close to the turbines. There were scattered reports that raptors, including golden eagles, were being found dead in existing wind farms. Noise and visual blight were also among the complaints being lodged against the wind farmers.

It was this same year that the voters of Sacramento elected to close the Rancho Seco nuclear power plant. A year later, the Sacramento Municipal Utility District, or SMUD, evaluated bids from power suppliers, and the Kenetech wind farm proposed for Solano County's Montezuma Hills made the final cut. SMUD would be the first customer to commit to Kenetech's new wind turbine, signaling that utilities, both public and private, were indeed becoming larger players in the unfolding story of wind power.

And it was in 1990 that Dale Osborn, whose square haircut and conservative clothes are anything but hip, declared that the wind industry "can't afford to look like Governor Moonbeam's children. This is not a ponytail industry." (A fired up Randy Tinkerman published a response in *Windpower Monthly*, an international wind power magazine, charging that Osborn's comments were a disservice to the U.S. wind industry. Without Brown, wrote Tinkerman, there would be no domestic wind industry, and noting that "No new technology grows without the dedication of idealists—many of which may have, at one point or another, sported a pony-tail.")

But the larger import of Osborn's declaration was that wind power now had to compete on a new playing field. The loss of state and federal tax credits and steep drops in fossil fuel prices dealt severe financial blows to many wind farmers already holding on by their fingernails in the late '80s and early '90s. Kenetech continued to reap profits thanks to lucrative power supply contracts with utilities that it and other outfits such as Zond had won in previous years, and only now were developing. But many other companies began to falter. Between 1981 and 1985 sales of wind-generated electricity rose from $21 million to $748 million, but by 1988 had dropped to $67 million.

By 1992, much of the industry was on the ropes. In that year, what was left of the wind industry received a shot in the arm when Congress passed the Energy Policy Act, which began deregulating the last remaining monopoly in America. Not only did this new law begin the process of opening up the nation's transmission grid to the buying and selling of electricity across the previously protected service boundaries of the various utility monopolies, but it also included a new tax credit for wind power. But unlike the investment tax credits engineered by Alvin Duskin and Ty Cashman in the 1980s, the new credit of the 1990s was granted only on the basis of actual energy production. Only those wind turbines that actually produced electricity would receive a subsidy of 1.5 cents per kWh—an amount that, it was reasoned, would make clean power generated from the wind more cost competitive with fossil fuels and nuclear power.

Ensuring that all buyers and sellers would pay the same price for transmitting electricity across the AC grid greatly improved the economics of delivering power from rural wind suppliers to the urban centers of demand for electricity. Or so the theory went. The new buzzwords in the electricity industry were "nondiscriminatory transmission access" for all, even wind farmers, the renegades who had originally fought against expansion of the now-ubiquitous electric grid.

◆

If many in the wind industry felt that they now had to compete on a level playing field, others began to see a double standard emerging. In fact, 1992 was a major turning point for wind farmers in a surprising new respect: The perceived negative environmental impacts of wind power suddenly became a hot button topic. Environmentalists, who had seemed to be a logical ally of the wind industry, were now the ones asking tough questions. Some went so far as to rally their troops to halt development of wind power. The cause of the public uproar was a study of the Altamont Pass performed for the California Energy Commission. It revealed potentially alarming levels of "strikes" or bird casualties from collisions with wind turbines. Using admittedly uncertain statistical models, the study estimated that a 6,500 turbine Altamont megafarm, stretching over 80 square miles of treeless rounds of hills, killed golden eagles at a rate of 39 per year. The report claimed most of the victims are chopped up by turbine blades or electrocuted by nearby power lines. This was not the first time that birds and other environmental issues would stop wind power development. But the fact that the report was published by a state agency that had been at the forefront of promoting this renewable technology (the same agency from which Duskin wangled the wind data for the Altamont Pass and for which Thomas, the Jung disciple, worked) gave more than a few wind power developers a new reason to fret.

If Cynthia Struzik had her choice, she would have shut down the Altamont Pass wind farm. The reason: illegal bird deaths. Struzik, a special agent with the U.S. Fish and Wildlife Service, the primary federal agency with statutory authority on the avian mortality issue, caused quite a stir at a meeting called by the California Energy Commission in December that year. Attendees included utilities, wind companies, private landowners with wind farms on their property, and a few environmentalists. The meeting was one in a long series devoted to the avian mortality issue, and most presenters spoke in the dry tones of academia or praised the economic virtues of wind power. When the audience was asked to comment on what number of bird strikes would be acceptable, everyone hemmed and hawed; even the environmentalists said they did not know.

But Struzik slowly strode up to the board in the front of the room and wrote "Acceptable Mortality Rate is ZERO." "The law is the law," she said. In fact, if taken at face value, the Bald and Golden Eagle Protection Act and Migratory Bird Treaty Act, forerunners of today's Endangered Species Act, could make criminals of any wind company operating a turbine that resulted in a single fatality of virtually any bird, including crows and other common feathered friends. Struzik then made an oblique reference to another, broader standard when she wrote "Thou Shalt Not Kill" on the board as well.

"I have twelve golden eagle carcasses sitting in the freezer right now," Struzik once told me, in an intensely angry voice. Wind power executives had reason to respect the strength of her convictions. In February 1993, for example, she expressed serious concerns over Kenetech's plans to study how its then new, and larger, wind turbine would affect bird populations in the Altamont. Rich Ferguson, the national energy chair of the Sierra Club, argued that such tests would lead to fewer deaths of eagles in the long run, but Struzik countered, "While the study may yield information, the study will not prevent migratory birds and eagles from being killed." Struzik successfully made the case that any installations of Kenetech's new and bigger machine should be matched with equivalent reductions in its fleet of existing turbines so as not to increase the probability of bird strikes. The rationale was that there should be no statistical increase in rotor swept area, since any increase would correlate with increased risks to protected species of birds in the Altamont. According to one estimate, Struzik's protests caused a reduction in approved new turbine installations and corresponding loss in energy production that added up to $10 million.

Ponytailed Tinkerman was the first to endure the wrath of environmentalists in the Altamont Pass in 1986. He is still extremely bitter about it. He had proposed to build an additional wind farm with Howden near the caves and hot springs he was so fond of. For the first time, environmental groups rose up in opposition, lumping Tinkerman's wind project with far more ordinary forms of industrial development. Birds of prey, such as the golden eagles that frequent the bare Altamont ridges; the San Joaquin kit fox; a rare species of brine shrimp; and species of plants all were listed as reasons why the Sierra Club and a local group called the Greenbelt Congress (which later changed its name to Greenbelt Alliance) opposed the 107-turbine project. The environmental impact report predicted that curiosity seekers would be drawn to the caves by the access roads the wind farm would require. An increase of visitors could only harm the Native American petroglyphs. Ironically enough, this environmental analysis was more concerned about kit fox dens than raptor deaths. The Environmental Impact Report postulated that the vibrations and soil disturbances associated with a wind farm might negatively affect newborn foxes. A proposed quarter- to half-mile buffer between turbines and eagle nests would hold disturbances to the local eagle population to an "acceptable minimum."

Yet another item that came up in the environmental impact report was a few clusters of Palm Oak, an evergreen shrub that happened to be native to the San Luis Obispo area several hundred miles to the south. The oaks, whose existence here can no doubt be traced to the wind, reside near the caves, as do vernal pools populated with brine shrimp for a brief period in spring. "Cows are allowed to trample all over the palm oak and the shrimp, but it is dangerous to place a wind turbine some 300 feet away," said Tinkerman, shaking his head in disgust. "I wanted to try and prove that a wind plant and sensitive environmental areas could co-exist. In fact, I argued that placing wind turbines here would offer greater protection to the caves, oaks, and shrimp than the status quo," he added. The wind farmers working in the field would be a presence that would deter trespassers from doing any harm to the delicate drawings. Tinkerman eventually succeeded in getting the permits to build the new wind farm, "but by that time Howden had already walked away from the wind business," he mumbled as he looked away.

Perhaps James Dehlsen had the most frustrating experience of a wind power developer dealing with environmentalists and birds. Zond had proposed to build a 77 MW wind farm to serve Los Angeles, a city that had become a symbol of what was wrong with our nation's air. The farm would be a great example of how clean energy sources could directly respond to growing concerns about smog and human health in the region. Dehlsen discovered that not everyone shares his enthusiasm for wind farms. The Audubon Society and Sierra Club express support of renewables such as wind power in their organization charters. Yet Dehlsen witnessed firsthand how well-heeled land speculators enlisting these former allies of the renewable energy lobby could fund and orchestrate a sophisticated political campaign to negate the virtues of wind energy.

In 1989, Zond proposed to develop a wind farm in the town of Gorman, an isolated community located in a sparsely populated region that straddles the boundaries of Los Angeles and Kern counties. Opposition galvanized on the issues of impacts on land values and loss of an endangered species, the California condor, a large raptor that was about to be released from the Los Angeles and San Diego zoos. Zond, which had sunk close to $1 million into the venture, had offered numerous concessions to environmentalists. It would install radio telemetry monitors that would emit signals when a tagged condor came nearby, triggering a shutdown of the wind turbines until the birds had passed through. A special fee per turbine would create a fund for local economic development. A 600-acre wildflower preserve was put on the table in response to complaints from some local citizens that the wind farm might disrupt their enjoyment of spring wild flower displays.

But the Tejon Ranch Co., whose owners include the Chandler family (which at the time owned the *Los Angeles Times*), was worried that the turbines might affect plans for posh condos in the area. It funded an instant organization called "Save our Mountain," which gathered 1,000 signatures in opposition to the Zond wind farm. According to documents obtained by the Kern Wind Energy Association, the Tejon Ranch Co. wanted to develop 230,000 acres for 10,000 new homes and a dude ranch and to lease additional land for mining. Of course, just over the hill, operating without a federal permit, a cement kiln was burning toxic waste as fuel, spewing noxious chemicals into the air. Enraged locals complained of strange sicknesses, and the small downwind town of Rosamond had one of the worst cancer rates in the state; many there thought the kiln had something to do with it. Residues from the kiln covered 1,000 acres, some of which sit on top of the region's aquifer. Also in the vicinity were an oil pipeline, microwave towers, and a colossal water aqueduct—hardly a place one would consider a pristine environment worth preserving at all costs.

One of the few public interest groups to back the Zond project was Washington, DC, based Public Citizen, the group founded by Ralph Nader. "The risks it poses to the environment are outweighed by the environmental benefits of wind energy," said Nancy Rader of the organization:

> The Gorman project should also be compared to the far more dangerous energy supplies that wind energy displaces. For example, it is instructive to note that at least 57 eagles and 25,186 other birds were killed in the recent Exxon Valdez 240,000 gallon oil spill alone. The Gorman wind farm, over its 30-year life, will offset fossil fuel generation equivalent to that produced by 10 million barrels of oil. . . . Global warming threatens many species which may be unable to survive even a gradual change in climate—a virtual certainty unless reliance on fossil fuels is greatly reduced within the next decade. In the worst case, whole ecosystems may disappear as a result of rising tides and shifting weather patterns.

Still, the influential Tejon Ranch Co. successfully publicized the condor issue, which influenced key members of the Sierra Club against Zond. At the time, seven Andean condors (considered the "ecological equivalents" to the endangered California condor) had been released into the wild to test the hospitality of the environment before the 25 to 30 captive California condors were to be released several years later. At the time of the debate over the Zond project, one of the Andean birds had already been killed when it struck a power pole. Another was returned to captivity after being unable to cope with the wild. The collision with the utility pole—though not in any way related to wind power—did underscore in the minds of more than a few that manmade structures, no matter how beneficial to the larger environment, could be lethal.

Typical of the propaganda disseminated was this statement by the Sierra Club's Les Reid:

> It is totally unacceptable for persons living in the South Coast Air Basin to attempt to shift even a part of the burden of our failure to adequately deal with air pollution to high wind areas near Gorman, in light of the environmental costs. To seriously impact the incredible beauty of the wildflower display, and to occupy a traditional flight corridor for not only condors but other scavengers (turkey vultures), raptors (e.g., golden eagles—"species of special concern")—all for a minuscule addition to our gross energy use, and of course a sizable amount of profit for the wind developer, is an unacceptable trade-off.

Though the local Los Angeles chapter endorsed the project, the state Sierra Club board overruled local members and helped convince the Los Angeles County Board of Supervisors to deny an operations permit for Zond. The Audubon Society was also enlisted in the effort to cancel the new wind farm. It was later revealed that Tejon funded Audubon biologists to hound Zond at public hearings. Ironically, Dr. Michael Wallace, a scientist with the Los Angeles Zoo Condor Captive Propagation Program, quietly supported the Gorman wind project because his "heart was with Zond." In a confidential memo, Wallace commented that he was upset that the condors were "abused for political purposes." From his perspective, "Tejon Ranch wants to develop condominiums and they think the wind turbines will be aesthetically unpleasing to potential buyers." He went on to lament the proposed development of condominiums because he felt they would affect the condor's chances of survival by taking away important habitat.

An exasperated Dehlsen, chairman of the board for Zond, complained at the time that his firm's wind farm would "generate energy and reduce air pollution. The public knows that and can't understand why a simple wind farm can't go forward in Los Angeles."

The day after the Gorman project was rejected, Dehlsen almost sold Zond to his archrival Kenetech. At least that's what Osborn claims. Osborn had just started his tenure as president of Kenetech, and like anyone who is anybody in the wind business, he had just spent a night at Dehlsen's ranch in Tehachapi. The night before the County Board of Supervisors' vote, Dehlsen had seemed supremely confident that the project would go forward. But the day after the rejection, he was despondent. "We could have bought our chief competitor for $1 million," Osborn quipped.

Part III

Chapter 18

The Windsmiths of Montezuma

Work is prayer. Work is also stink. Therefore, stink is prayer.
—Aldous Huxley

Bill Graham is climbing an 80-foot aluminum lattice tower, wind shooting through his thin brown hair and pushing against his beer belly as another February storm rolls in from the Pacific. At 40 mph, the wind is barreling up the narrow Carquinez Straits, where salty waters from the Pacific mingle with fresh water from the Sacramento River. The steady stream of wind follows the stream of water below and howls toward a heap of soil that has been christened the Montezuma Hills. It's getting near quitting time and Graham, on the job for just four months in 1994, shows a few white knuckles as the tower sways with each bang of wind. A "windsmith," Graham needs to tighten some bolts so that the propellers on the very top of the tower he just climbed can start spinning and converting zinging molecules that comprise a gust of wind into electricity.

Graham, while a fan of really loud rock 'n' roll, is not *the* Bill Graham of San Francisco rock biz promotion fame. That Bill died in 1991 when his helicopter collided with a utility transmission tower about 35 miles to the west of this spot. This Bill hails from the nearby Sacramento Delta town of Rio Vista, and he seemed to be quite happy with his new line of work, in the employ of Kenetech, the largest wind farmer in the world. "Best thing about this job is that you're out in the middle of nowhere, out here in the elements," he enthused.

When up on top of a tower, you can pick your horizon, he told me. To the immediate southwest looms Mt. Diablo, a windswept dome local Native American tribes considered the center of the world. Directly south, on a clear day, you might catch a glimpse of the Altamont Pass, alive with over 6,000 wind turbines feeding off warm, dry winds that cling to the surface of mountain ridges. To the immediate east is the Sacramento River, meandering its way west to the salt of the Pacific, but not before its waters seep into the many fat fingers of the Delta. The white crowns of the Sierra Nevada rise in the distant east. Further north, the snake-infested Sutter Buttes, the smallest mountain range in the world, break up the flatness of the surrounding valley floor. The sun slowly crawls across the big open blue canvas and by the end of the day Graham is ready to collapse. His wife told me on workdays he's often snoring before 9 o'clock. The constant exposure to the wind all day long just saps the living daylights out of him.

Graham and a buddy are responsible for 200 or so of the over 600 wind turbines in this wind farm. In the off-season, winter and early spring, Graham spends an hour on each tower applying some "PM" (preventive maintenance) consisting mostly of oil and grease work and the tightening of screws that are loosened by all that vibration from the wind. This wind farm produces two-thirds of its electricity in the six middle months of the year. April Fool's Day is the deadline to get these wind turbines ready. All machines here in the Montezuma Hills have to be in primo condition. "If all goes well, we're sitting on our butts in the summer, or throwing darts or Frisbee," Graham said, nursing a Budweiser and giving the first hint of a smile.

Standing below the large steel structures, it is hard to imagine climbing up and down, step by step, five or six or seven of these in one day. In the Altamont, Kenetech has towers as tall as 140 feet. "They find out pretty quick if you are going to be able to do it, or not," Graham said, referring to the site managers. "If you think about falling, you can talk yourself into getting the shakes," he adds, pretending to vibrate from an intense wind. The wind of the Altamont can be brutal, even violent. "We have much softer winds here," noted Graham, as he gazed out toward the smooth ridges and gentle rolls of the lazy Montezuma hills. "I like it out here in this little corner of nowhere," he confessed. "It's far away from the madding crowd." The norm for Graham and his cronies is to quit clinging to the towers when winds get to 45 mph. At that speed "the wind can knock you off-balance. And when the tower is moving, you can lose your tools." The 60-foot towers, having only three legs, actually vibrate more than the 80-footers, which stand on four legs.

Though these windsmiths could easily qualify as rednecks, their work requires more than just screwdrivers and wrenches. Some fundamental computer skills are also required. Seemingly isolated in the outback, this crew in the Montezuma Hills is actually wired directly into the company's headquarters in nearby Livermore, where the command center for the hi-tech end of Kenetech's wind farming operation is located in a rather ordinary industrial park. Inside the complex, five computer modules create a semi-circle. The advent of cheap and powerful computer systems since 1990 had a tremendous effect on operations like Kenetech's. Thanks to custom software, three to four people can optimize the operation of the company's 4,000 or so California turbines, picking and choosing which turbines to run in order to generate the most power the most cost effectively. At times, the wind may blow from such a direction that individual turbines suck up the kinetic energy before it can get to the next turbine in a turbine string. It might make sense to turn on only every other turbine, for example, if the wind is blowing parallel to a ridge that is densely packed with turbines. (Most other wind farmers during the early California years just allowed their turbines to operate on autopilot, though today sophisticated computer controllers are common.)

Kenetech rode to the top of the wind industry on the backs of windsmiths like Graham. No other wind developer put so much of its resources into integrating the real-time computer controls of each individual turbine with immediate, on-the-ground technical assistance. Without such constant attention, the performance of Kenetech's machines would have ranked much closer to that of Fayette's machines. The windsmiths who worked for Kenetech became a key part of the company's formula for success. They provided invaluable feedback about how each component fared under the very specific wind conditions and quirks of wind fuel supply at any one turbine site. It was this level of attention to detail that allowed the company to keep as many turbines on-line as it did, wringing millions of kilowatt hours of electricity out of the wind. Since it was not unheard of for parts of a machine to literally just blow away, Kenetech developed a cadre of windsmiths who became crack operation and maintenance hands.

By 1994, Kenetech had more than 600 turbines on the Montezuma Hills, but the vast majority of Kenetech's turbines—3,500 100 kW machines—were located in the Altamont. But it also had 75 turbines planted in the San Gorgonio Pass near Palm Springs, about 300 miles to the south. In the old days, the company monitored all of its machines on-site with Apple II computers; 50 turbines were plugged into each Apple. In 1986, the company moved its operations to its current central location and switched to PCs driving custom software that continually tinkers with the operation of the turbine fleet. "Each day we arrive about 6 o'clock in the morning and check the computer printouts," Graham reported. The printouts present the latest info on all operational aspects of each turbine in the wind farm since quitting time the previous day. Each turbine reports on its performance every two minutes, updating the screen. If an operator wants to focus on one individual turbine, data can be updated every two seconds. Most of the time the computer spits out its routine stuff. Once, however, some curious outages were explained as sabotage: Vandals had cut the power cables with hacksaws for no known reason.

State-of-the-art wind farming combines the on-the-ground mechanical know-how of windsmiths like Graham with hi-tech computer technologies to convert the raw energy of the wind into steady pulses of electricity. These windsmiths know all about the weather. Like conventional farmers, wind farmers notice the subtle scents of the rare oncoming thunderstorm, the thick taste of dry August heat swallowing you in its hot breath, the cool soothing breezes that flow in from the Pacific in early summer evenings. As more traditional forms of farming the land are lost to new subdivisions and countless low-paying jobs at shopping malls, there will be at least one profession that still requires people to work out in the elements.

"It's just so peaceful when you are up on top of the tower," remarked John Opris, who has been a windsmith since 1987, working his first two years at the Altamont Pass before moving over here to the Montezuma Hills.

"You are just like a farmer watching the weather to see how it is going to affect your crop," said Opris, who sports a crew cut and wears those glasses that darken in bright light. "Your turbine's performance is directly related to the weather on a minute-by-minute, hour-by-hour, day-by-day, basis," he said. "The highest height of the landscape here in the Montezuma Hills is 300 feet—maximum," he continued, noting that weather conditions are usually quite tame. Since the Altamont has elevations as high as 2,000 feet, the winds and weather there are more dramatic. "In the Altamont we had some unique frosts and lots of weird cloud cover. In the morning sometimes, your co-worker might climb up a tower right into the clouds. It's sort of like he's climbing up a beanstalk or something. Sometimes the fog is so heavy that you can't see anything when you are up on a tower. Other times, it is clear as day and you can watch hawks do acrobatics all day long."

Opris actually claims to hate the wind. "When the wind is concentrated—like in southern California when the Santa Anas blow—life is miserable. Here at work, there are times when you've been up on the towers too long and you get almost an anxiety attack and you just have to come down." On average, Opris calculated that windsmiths working in the Montezuma Hills climb about three miles of tower ladders per year. And while he may hate the wind, Opris loves his job. "It feels good working for a good cause. And nobody's looking over your shoulder. You can play tunes, you can watch the wildlife. When you're up in the towers, you can see smokestacks from the industrial parks in the Delta, stinking up the place, and here we are generating electricity without any pollution."

The Montezuma Hills are surrounded by amber waves of grain. This new kind of farming fits right in with the traditional agricultural trades. Collecting electricity from the wind takes a lot of old-fashioned muscle, everyday mechanics, and common sense at this end of the wind farming operation. All but one of the guys (and one gal) who work for Kenetech at the Montezuma Hills are locals. Graham lived 10 minutes from work, in a modern house just a few blocks from his wife's parents, who reside in a Victorian house built in the late 1800s and which features a dozen bedrooms.

It was Kenetech's unique ability to combine the on-the-ground local knowledge of its windsmith technicians with high-tech tools like computers running customized software performance-enhancement programs that enabled them to succeed where others had failed with machines that required so much ongoing maintenance. The windsmiths working for Kenetech shared their intimate knowledge of the local wind resource, contributing immensely to impressive production numbers at a time when most of its rivals began to rely on newer, and much larger machines. The key question was whether Kenetech would lose sight of this local knowledge in its rush to grow, to design to cost, to please the moneylenders. As the company expanded its wind power empire, the focus was to meet aggressive turbine sales goals. Wind turbines were now being installed in quite varied terrain. Windsmiths were starting from the ground floor in these new wind farm locations. The company's core strength—the link between computerized management and windsmiths in the field—was being tested in new ways.

Chapter 19

From Boom to Bust?

There is much to be said for failure. It is more interesting than success.
—Max Beerbohm

Robert Lynette, a tiny man whose face with its chiseled features would look great on a dollar bill, knows all about the financial balancing act required for wind farming to work. He once managed the nation's largest wind power consulting firm. "I had 25 people working for me at one time," he told me, harking to a time when his firm tried to keep an infant industry honest. It was Lynette's analysis of the wind turbine market in the early years that convinced Merrill Lynch to move forward with all of the back-to-back USW limited partnerships in the Altamont Pass. His word carried more weight than perhaps any other wind industry analyst during the 1980s.

Lynette took a big gamble in the 1990s. He traded in his unbiased consultant cap for the hat of a company executive with FloWind, a San Rafael-based wind turbine developer and manufacturer. FloWind had made its mark in California with its vertical-axis turbine, and was the only major California company to build a business with such an "eggbeater" design. But, in an effort to diversify FloWind's portfolio of products, Lynette introduced yet another horizontal-axis, two-bladed, downwind American-style lightweight turbine into the market. This Advanced Wind Turbine (AWT) was one of the few DOE designs to ever enter full-scale commercial production. Many in the industry viewed this as a very odd development. None of the DOE machines boasted great performance statistics—and Lynette's consulting firm had made more than a few bucks documenting the kinds of problems these, and the rest of the industry's turbines, had.

Lynette's critics quieted a bit when it was reported that FloWind landed a major coup with a sale of almost $500 million of his AWT machines to India, which had suddenly become a hot new wind power market. The first shipment, of $12 to $13 million, helped bail out the company, whose continued existence was an extremely delicate balancing act. Nevertheless, instead of continuing to refine the AWT, which was also picked for a new wind farm in Washington State, upper management remained obsessed with validating the vertical-axis machines. All of its spare cash went into testing a new, extremely tall and skinny vertical-axis prototype turbine—design characteristics that Lynette complained "violated physics."

Lynette claims that the principal shortcoming of the vertical-axis design is that the blades are too low to the ground, and so reap only the lower grades of the wind fuel. "You give up 20 percent right there with a vertical-axis machine because of the lower velocity of wind so low to the ground," he said. The new vertical-axis turbine was taller and narrower in order to capture the available energy at higher elevations, but this made another primary problem with vertical-axis turbines become even more pronounced: severe stress on the blades from multi-directional wind shear. Winds engage the bottom, middle, and top of these vertical-axis blades at different speeds. These wind differentials place severe stress on the entire machine, introducing vibrations that can provoke the kind of catastrophic failures that destroyed the Alcoa vertical-axis machine at the first Palm Springs wind power conference back in 1981. Making the rotor swept area even taller and narrower than previous designs, and adding a third blade, increased the challenge to the turbine's long-term integrity. In other words, the design was doomed. Despite these logical arguments against the new design, "Adam and Eve," as the two vertical-axis prototype turbines were called, soaked up virtually all of FloWind's spare cash, pushing the company to its fiscal limits.

Lynette didn't want to stick around, because he thought Adam and Eve were dangerous. Sure enough, they ultimately self-destructed, as did FloWind. "There went three years of my career," said Lynette. The official story was this: FloWind thought it had made a sale of over 400 AWT turbines to India. But the vagaries of doing business in India, where business customs were apparently not understood by U.S. companies, led FloWind to deliver turbines for which the Indian customers ultimately refused to pay. According to Lynette, in a fit of desperation, FloWind manufactured hundreds of wind turbines "based on a person's word" and a purchase order. The company invested $5 million in these AWT turbines without ever having a letter of credit or other guarantee of payment in hand. The decision to move forward with fulfilling the order was, in Lynette's view, a foolish act of desperation made necessary by upper management's preoccupation with Adam and Eve.

◆

One company that survived—though it never installed a commercial wind project in California with the help of tax credits—was the Wind Harvest Company, the tiny firm with a vertical-axis turbine funded by associates united solely by their beliefs in Jung. Once the wind rush was over, Bob Thomas left the California Energy Commission and returned to the task of transforming his red star dream into a real piece of power-generating technology. Unlike the latest FloWind vertical-axis design, the Wind Harvest turbine is a squat device that looks more like the upright solar panels mounted on space satellites than a wind turbine. Unlike other vertical-axis turbines, the Windstar features fifteen blades grouped into five modules—hearkening back to the five-pointed star of Thomas's original dream. Support arms from the top and bottom reach out from the center to create the five spokes, each spoke attached to three blades stacked on top of each other to form a single vertical plane. The blades are fixed pitch. The machines are actually wider than they are tall: 50 feet in diameter by 46 feet tall. Unlike the FloWind machines, the Windstar is self-starting. It is among the quietest of wind turbines, since the tip speed of its blades is only about half that of the traditional Danish horizontal turbines on the market in the 1980s.

The Wind Harvest Company scored its first significant success when it was picked by the Communist Chinese government to install one of the first commercial wind farms in that country. The selection caught everyone completely off guard. Unaccustomed to the reality of dealing with an actual customer, Wind Harvest went out of its way to not lose out on this promising opportunity. To grease the wheels of the proposed purchase of ten 25 kW Windstar turbines, George Wagner arranged for a layover for visiting Chinese officials in Hawaii, so that the large, rotund Chinese government officials "could see girls in bikinis." He added, "You have to remember, we did the layover for guys who lived in a country where if you were caught with a *Hustler* magazine, you would be shot." There was talk of installing a Windstar near the Great Wall, a symbolic gesture that the Jungians reveled in. But the promise of long-term sales never materialized. Wagner thinks the Chinese ripped off the design and just started manufacturing their own.

Still, other Jungian investors kept Wind Harvest alive. They include Sir Laurens van der Post, who was one of Wagner's childhood heroes and the subject of a film he produced. Van der Post, who was quite close to Jung, didn't give a hoot about the environment when he wrote a check to Wagner. "His sole reason for the financial help was that he wanted to see the dream become concrete in the real world," Wagner said. Van der Post's credibility made Wagner's job easier, especially with members of the rock band The Police. One of Wagner's most embarrassing incidents was standing up Sting, the lead singer of The Police, at an airport. The Police had just released *Synchronicity,* a huge international hit, and all of the band members had become bloody rich. Sting was reportedly interested in the Windstar because his wife had heard about it while attending a lecture about Jung. Wagner got stuck in traffic and showed up two hours late. Though Wagner and Sting chatted quite a few times after that, Wagner never had a chance to ask him directly for money. But Andy Summers, the band's guitarist, and a van der Post groupie, eventually chipped in. Another Jung connection brought Rob Lowe into the Wind Harvest circle. His mother was into Jung. Lowe even filmed a commercial promoting the Windstar. The tape was on its way to Ted Turner, who expressed great interest, when Lowe became embroiled in a New Orleans sex scandal. "Just our luck," chortled Wagner.

The Wind Harvest Company's biggest break came in 1992 when it secured a $3.5 million power sales contract in a competitive solicitation issued by utilities in the United Kingdom. Through its contacts within its network of Jungians, Wind Harvest was granted wind rights to a ridge on the Queen's property. The turbines were installed, but Wind Harvest had fallen $100,000 short in funds needed to connect them to the all important UK grid. Thomas, walking along the ridge of the installation, was contemplating where the firm would get the money, when it dawned on him. The ridge, the turbines, this landscape had been in the dream, the one that led to the creation of Wind Harvest in the first place. To Thomas, this meant the wind farm project was destiny.

Nonetheless, the turbines were installed but couldn't be energized. The money had run out. Even a proposed joint venture project with Zond, one of the strongest companies in the industry, fell apart at the last minute because of Zond's tenuous financial health.

Wind Harvest came within one month of total collapse in 1994, when a private investor stepped up to the plate. His name? Ty Cashman. Wagner first met Cashman here at Dipsea's, a Marin County café where I conducted one of my several interviews with Wagner. He told me he was watching the credits on a film about Tibet when Cashman's name appeared. Thomas had told Wagner that Cashman, having long moved on from his stint in Sacramento with the Brown administration, was teaching philosophy in Minnesota. Turned out he had just returned from the Midwest. Wagner tracked him down, and the two met at this same cafe and chatted for four hours. Shortly thereafter, Cashman, now heading up the Solar Economy Institute out of his cozy Mill Valley hideaway, handed Wagner an envelope. It contained a check for $110,000. The first turbines were then hooked up to supply clean electricity for the English aristocracy. Once again, an unexpected check kept hopes of the harvest alive.

Then the bottom fell out again. The only Fortune 500 company to invest in the Windstar, to the tune of $300,000, was the Morrison Knudsen company. Not only did Morrison Knudsen go belly-up, but then Sam Francis, the painter who had invested over $1 million to get the company off the ground, passed away, too. Thomas went on unemployment. A good friend of mine, Wagner told me, helped cover some phone bills so that the company could stay afloat a few years back. Last time I talked to George Wagner, he was hitting the phones again, trying to rustle up some more big change from his various lists of the rich people of this world. He confessed, "This thing is driving my wife nuts. We've got to get some regular revenue going." He explained that his spouse was a high-powered, high-paid attorney. His own psychiatric business had suffered immeasurably from the obsession with the Wind Harvest cause.

But things were looking up, he assured me, after breakfast, as we parted ways in the rain.

◆

Fayette, of course, had fallen victim to Eckland's strategic mistakes, chiefly the poor sites chosen for his turbines because of his obsession with the Tesla substation in the Altamont Pass. The IRS was stepping up pressure on the embattled company to sell its assets by the end of 1987 in order to pay off its multi-million dollar debt of back taxes. Despite the consistently poor performance of its wind turbines, the assets of the company were snatched up that year by the Wolverine Power Corp., a company headed up by John Kuhns, a darling of Wall Street and one of the nation's best-known alternative energy entrepreneurs. The price was allegedly $8 million, about the same amount Fayette owed the IRS, for all of the company's turbines, infrastructure, land leases, and power sales contracts, the last representing 650 MW. The firm had developed only 150 MW, but its contracts, unlike those of Kenetech and other wind developers, paid out prices for power that were pegged to fossil fuel prices. Unrest in the Middle East was once again fueling speculation that oil and natural gas prices would increase, and some predicted that the unused power sales contracts could be valuable in the not so distant future. But they were dreaming.

In 1982 Kuhns, a slick multi-millionaire and smooth operator, helped start up a much ballyhooed company called Catalyst that was publicly traded. Within three years, Catalyst was worth $9 million and Kuhns had engineered a number of acquisitions that created a 260 percent jump in revenues and an 830 percent increase in profit. But those who did business with Kuhns were often not very happy. For example, his first independent power project, a small hydroelectric power plant developed for the Yolo County Water District, in California, was secretly sold to T. Boone Pickens even though Catalyst still owed the water district close to $1 million.

Many wondered what Kuhns intended to do with the problem-plagued Fayette turbines. Eckland was quickly booted out, but soon was back on the company's board. He and Kuhns were buddies. They had both invested in a technology called the Kalina cycle, which was touted as boosting the efficiency of thermal power plants by 50 percent. Eckland claimed to have helped bring its Russian inventor, Alexander Kalina, to the United States and to have purchased the worldwide rights to the technology at one point. Indeed, Eckland maintained that he thought the long-term future of Fayette rested with this rather than wind technology.

Kuhns made noises about grand plans for Fayette shortly after the purchase. The company would diversify into wind turbine blade manufacturing and solar photovoltaics. Kuhns then purchased another small, pioneering wind turbine manufacturer, Vermont-based Northern Power Systems. Despite Kuhns's promises, nothing really happened. Operations and maintenance of Fayette's Altamont turbines was subcontracted to SeaWest, but that company was never paid. Kuhns further confounded observers when he purchased in Hawaii 14 Westinghouse MOD-2 650 kW machines, which never won any awards for performance. Rumor had it that Kuhns wanted to turn the Hawaii wind farm site into a tourist attraction. "All I was really after were future sites and the utility power purchase contracts," said Kuhns in trying to explain his purchase strategy. "I thought the turbines would maybe be sold off as scrap. After all, I did not purchase the patents on the Fayette technology—that still is in Eckland's hands."

Kuhns then folded Fayette under the umbrella of his new business venture, New World Power, and began aggressively pursuing overseas projects, lining them up in Costa Rica and the British Isles. One measure of his apparent rise in respectability was that he was chosen to deliver the keynote speech at the 1993 AWEA conference. The fundamental thrust of his remarks was that cheaper money was at least as important as the actual wind technology, if not more important. He called for the creation of a mutual fund targeted at raising investments for new wind and other renewable energy projects. Kuhns outlined an approach that seemed to echo Eckland's goal of reaching out to individual investors willing to take on a higher risk to get a fixed return on their investment. He argued that the high risk that institutional investors placed on wind energy was greater than the real risks of putting wind projects in the ground. This obstacle could be overcome, said Kuhns, by tapping new investment sources. According to one anonymous insider, Kuhns's approach had already amassed $7 to $15 million "by selling wind power to Joe Blows who know nothing about the technology."

Like Eckland before him, Kuhns was building what looked like a real company on paper. The reality was something less. The assets Kuhns was accumulating helped him to attract additional investment dollars, but they provided little in the way of operating revenue.

Kuhns wasn't making life for Zond any easier, either. Zond was still reeling from the rejection of the Gorman project near Los Angeles. Particularly frustrating was a potential deal in Chile, in 1993, on which Dehlsen had been working for months. Although Zond trimmed the project as much as it could, the developer would still need 6 cents per kWh to work. But suddenly the Chilean government informed Zond that the payment would be capped at 4.5 cents per kWh. Dehlsen later discovered that Kuhns had brought in a beautiful 21-year-old translator who knew virtually nothing about wind power and was therefore quite inept at explaining the complexity of the proposed transaction. Members of the Chilean Embassy said the episode was embarrassing. According to Dehlsen, New World Power took all the projects that Zond had rejected because the economics would not work. Still, these projects were trumpeted by New World Power as evidence of its growing presence in the global wind power market.

Chapter 20

Betting the Farm

> You don't need a weatherman to know which way the wind blows.
> —Bob Dylan

As the exceedingly generous investment tax credits that had jump-started the industry in the '80s faded into the sunset, to be replaced by far more limited production tax credits in the early '90s, wind farmers debated the long-term prospects for wind power. Since oil prices, which never got near Eckland's $100 per barrel forecasts, had also collapsed, the verdict was this: If the technology was to become a mainstream source of electricity, bigger and better wind turbines would need to be developed. They would have to be able to operate both at lower and higher wind speeds than existing machines in California's wind farms. PG&E and a handful of other utilities began to look at wind power as a technology they might want to invest in, not because of cost considerations—wind power was still more expensive than most fossil fuel options—but because of wind power's strategic value as a hedge against any future jump in fuel prices.

In 1987, the Electric Power Research Institute (EPRI) and USW co-sponsored an advanced turbine feasibility study. Virginia Power and the Bonneville Power Administration, the latter a huge federal agency that managed a series of hydro powerhouses in the Pacific Northwest, participated in the technical reviews. The fundamental goal of this feasibility study was to determine whether the incorporation of computerized power control, which would theoretically capture more power from the wind, could bring the cost of wind power down to 5 cents per kWh—the cost at which wind farming was deemed to be competitive with fossil fuels.

Kenetech went to work developing a new turbine, the 300 kW 33M-VS.

Although the machine represented a departure for Kenetech, in that, like the Danish machines, it faced into the wind, the most radical element in the new design was a variable-speed rotor. Most machines on the market at that time featured a constant-speed rotor. When winds exceeded a threshold (in the case of Kenetech's 100 kW machine, winds over 44 mph), the turbines were shut down. This protected the blades, but it also kept the machines from wringing electricity from the very winds that possessed the most kinetic energy. Since the variable-speed rotor can speed up or slow down, it was expected to boost power production by 10 percent. This ability to capture more energy at the high and low ends of wind speed conditions (in the language of wind industry insiders, the machine had a larger wind-speed "operating envelope") was thought to be a major factor in reducing the price of wind power. In addition, the variable-speed operation of the turbine generator offered utility owners something new. "The machine really has the potential to be a good neighbor on the power grid," commented J. Charles Smith, an EPRI consultant. "Even when the wind is not blowing and the turbine is idle, the power electronics could be switched onto the utility line to provide reactive power support for voltage control. By putting power back into the grid to stabilize voltage levels instead of just taking power away from the grid to make up for the intermittency of the wind, the Kenetech machine could transform wind power from a potential grid interconnection headache into something that provided value to a utility in the form of greater grid reliability." Kenetech touted this advance to its utility allies. The unique variable-speed power electronics component was Kenetech's new technological advantage, and the company claimed this new wind turbine offered operational value to a utility that went beyond its actual kilowatt-hour production.

Following on the heels of a surprisingly successful public stock offering by New World Power, which insiders considered a controversial player because of its purchase of the much maligned Fayette assets, Kenetech began preparing an Initial Public Offering (IPO) to raise funds to ramp up production of its 33M-VS wind turbine in 1993. This was the dawning of a new era for wind power, it seemed. There was no better time to go public, as the limited partnerships that had been put together by Merrill Lynch for Kenetech in the 1980s had run their course. It was time to gamble the future of the company on new technological breakthroughs that would help make wind power a widely acceptable way to generate electricity for utilities.

Kenetech's variable-speed rotor technology was not unique. Enercon, a firm started in the early 1980s to manufacture and market industrial inverters, had a variable-speed turbine from the beginning. The main application of these inverters was in variable-speed industrial drive systems used in robotics. But Enercon also introduced a wind turbine with a variable-speed rotor and fixed-pitch blades in 1984; its first commercial turbine, the E-32, was rated at 300 kW—the same size as the Kenetech machine. The company sold 189 of these units between 1989 and the end of 1993. One of the unique features of a successor Enercon design, the E-40, which New World Power had proposed to use in a 40 MW project in Big Springs, Texas, was its gearless drive system, made possible by advanced power electronics and other innovations. A gearless turbine makes no noise and requires no oil changes and virtually no gear replacements, translating into longer operating hours and lower costs.

In its 1993 IPO, Kenetech implied that it was the prime innovator of variable-speed rotors. Even shrewder, however, was language Kenetech put in its variable-speed rotor patent. The patent was written so broadly that it ultimately blocked Enercon from doing any projects at all in the United States. Under American patent law, the applicant is in charge of establishing what the legal world calls "prior art." In other countries, "it is up to the authorities to make that determination," sighed Mark Haller, the U.S. representative of Enercon. That's how Kenetech got away with the patent, even though Enercon employed a variable-speed component at least one year prior. Haller also blames the owner of Enercon, a farm boy whose hold on the variable-speed patent was compromised by trade secrets still owned by former engineering partners. New World Power took Kenetech to court to try to salvage the Big Springs project, but Kuhns relied on an attorney who had no fire in her belly to handle the case. "It certainly wasn't a knife fight," Haller deadpanned. His biggest lament was all the business he had generated for the Enercon wind turbine. "It [had been] such an easy sell following Kenetech because they were such arrogant pricks." Still, Kenetech had proved again it knew how to play hardball and come out on top.

Kenetech parted with the rest of the industry in other ways as well. It had told the whole world its new state-of-the-art 33M-VS wind turbine could produce electricity for 5 cents per kWh—at the time, the holy grail of wind power production. Indeed, it now maintained that the wind power industry no longer needed a helping hand in the form of tax breaks or government subsidies. The combination of power electronics, a variable-speed rotor, much larger generators, and much lighter weight all contributed enough cost savings to make wind power attractive on its own merits. The company wasn't going to ignore the production tax credits now offered by the federal government. Just the same, it made a point of distinguishing itself from other wind farming companies that continued to call for preferential treatment of wind and other renewable resources in light of their contributions to cleaner air. All Kenetech wanted was the so-called "level playing field."

An incident aptly captures Kenetech's attitude at the time. In 1992 DOE finally responded to complaints of struggling wind farmers about how it divvied up its budget to advance wind technology. For the first time, new funds were earmarked to co-fund demonstration projects for Zond and Carter turbines and to improve blade airfoil designs that would be of value to the entire industry. Kenetech raised a stink. In September 1993 Gerald Alderson, now Kenetech

CEO, sent a scathing letter to Congressman William Baker, who represented Kenetech's district. He questioned why DOE would develop programs "whose specific purpose is to spend taxpayers' money to create a competitor for us. We don't mind competition; we have it," he said, referring to companies based overseas, adding, "What we ask is a level playing field. As we see it, there is no possibility that a level playing field consists of the government funding the creation of competitors for us. . . . It is impossible for our shareholders to compete with funding from the federal government (the ultimate 'deep pocket')." Baker sent a letter to DOE asking Secretary of Energy Hazel O'Leary to re-examine the DOE program. Nonetheless, DOE funds flowed to installations of the first-generation Zond turbines in Texas and Vermont.

A widening gulf emerged between the rest of the industry and Kenetech, which came to be viewed as a bully, a company that didn't seem to have a problem adjusting to the self-centered business culture of the 1990s. But among the general public and in the halls of Congress, Kenetech was seen as the only credible company in the United States offering sophisticated wind farming technology.

◆

As of summer 1993, Kenetech had already spent $72 million developing its new turbine. It was hoping to manufacture 500 in 1994 and twice that number in 1995. The company acknowledged in its IPO that it needed $10 million of working capital to meet these aggressive production targets. The IPO was to be for one million shares, but the response was phenomenal. The firm sold six times that amount at prices ranging from $16.50 to $25.50. Buoyed by such a robust response from financial markets, the company announced that the majority of the $92.4 million it raised would be earmarked for the development, marketing, and production of its turbines. Some $25 million in senior secured notes would also be paid off.

Merrill Lynch was not the only underwriter of the IPO; Morgan Stanley & Co. and Smith Barney Shearson also got a piece of the action. While the rest of the wind industry faltered, Kenetech had become a hot investment opportunity again. Merrill Lynch predicted Kenetech's sales would increase 100-fold between 1992 and 1995. In a report entitled *The Answer Is Blowing in the Wind*, Morgan Stanley forecast that the value of Kenetech stock would rise by more than 50 percent within a year and that by 1995 earnings would reach $1.35 per share. Smith Barney also tooted Kenetech's horn, one analyst choosing it as the top stock pick of the year, predicting a $40 price per share within the next twelve months. The three largest shareholders—The Hillman Co. (36 percent), Allstate Insurance Co. (10 percent), and F.H. Prince & Co. (8 percent)—reflected the respectability wind power now commanded in financial markets.

Did Kenetech's performance merit its status as darling of the investment community?

The IPO's success was surprising because only 22 33M-VS prototype turbines were operating in the Altamont Pass at the time. Not a single commercial unit had been put in the ground. SMUD was among the first to commit to purchasing the new turbines, signing an agreement in February 1994. Claiming SMUD to be "the first utility in the country to build a commercial-scale wind farm," general manager Jan Schori commented, "Wind energy represents a zero-emission fuel source for utilities. But it wasn't within our grasp until a new generation of advanced design turbines retrieved more work out of wind at 40 percent of the cost of standard turbines."

Kenetech counted the SMUD project as 50 MW even though the terms of the deal called for a 5 MW pilot project to be evaluated before building out the remaining 45 MW. The company had another 169 MW in what the firm referred to as "executed" contracts and another 430 MW in "pending" deals. At this point, Kenetech had an installed base of 4,200 of its 100 kW machines.

The firm's profit was close to $10 million in 1991, but the following year it was only $2.8 million, and in spite of the investment pundits' predictions, things did not look good for 1993. As of July, weak winds had contributed to a dismal power production tally for the first half of the year—39 percent below average. On top of that, revenues derived from the power purchase contracts between Kenetech and PG&E were diminishing. Under terms of the contracts, wind mills in operation for 10 years would see energy prices move from Eckland's high oil price predictions to real prices, which were low and declining. This dropped revenues another 12 percent compared with the previous year. The company acknowledged in 1993 that it would not produce any profit at all. It ended up in the red for the first time since 1982.

But demand for wind power remained strong. In April 1994 Kenetech again went public, this time selling preferred stock and raising still more funds to help meet the burgeoning demand for its cutting edge product.

◆

At about this same time, Kenetech had become engaged in an all-out war among wind companies. The first new utility power purchase contracts in California in well over a decade were up for grabs. Instead of offering plump tax credits to wind farmers and other renewable energy developers, state regulators would quantify the economic value of air pollution avoided by reliance on clean energy sources. These economic values, assigned to each type of air pollutant, were added to the price renewable energy developers bid, and tilted any analysis of cost effectiveness toward clean power plants.

The CEC and California Public Utilities Commission (CPUC) spent years debating these values, as well as a series of complex rules that were to govern this introduction of "competition" into the selection of new power sources. Bureaucratic wrangling stretched formulation of the Biennial Resource Plan Update (commonly referred to as the BRPU) from two years to eight years (the proceedings to develop the auction began in 1988 and those to reach final resolution stretched beyond 1996). The BRPU required the state's three investor-owned utilities to add 1,358 MW of new power to their systems. Each utility had to calculate what it would cost for it to build various power plants the CPUC deemed to be the least expensive, then wind farmers and all other private company bidders, including those who wished to burn fossil fuels, would bid against these utility proxies.

The results of the auction shocked everyone. Independent energy producers submitted bids whose prices were 17 to 44 percent below what utilities said it would cost to build new fossil fuel power plants. The BRPU sparked intense competition among wind companies; they offered the lowest prices ever for renewable energy in the United States. "The most interesting thing to come out of the BRPU was the simple fact that every single bid in the Edison service territory—even the wind power bids—beat Edison's proposed natural gas plant repowering. If you told me that a year ago, I would have told you you were crazy. But that's a fact," said Jan Smutny-Jones, executive director of Independent Energy Producers, the group that Hamrin started up after experiencing the frustration of no-win negotiations with Southern California Edison. According to Smutny-Jones's math, rate payers in Edison's service territory would save $500 million thanks to BRPU projects, including 585 MW of new wind projects to augment the 1,700 MW of wind capacity already in the ground.

The lion's share of the 585 MW went to Kenetech and SeaWest, some of whose bids nonetheless would have required their wind farms to operate at more than three times realistic production levels at energy prices near zero. Some final prices were actually *negative* after the 1.5 cent per kWh federal production tax credit was factored in. These bids were strategic responses to the way the BRPU bids were structured and scored. For example, a bid might include a price for the energy delivered into the grid of only .5 cents per kWh. When the 1.5 cent per kWh energy production tax credit was factored in, the cost of the electricity on paper was –1 cent per kWh. To make up for this low-balled energy number, the bids included projections for capacity (the actual proportion of the time the wind farm would be operating and delivering power to the grid) to 60, 70, even 80 percent. In the real world, California wind farms usually operated somewhere under 30 percent of the time. Multiplying the low-balled energy price by the hugely inflated capacity figures resulted in revenues that would enable the bidder to turn a profit, but still bid a price lower than what it would cost a utility to build a new fossil fuel plant. Whether these bids were legitimate or not was a matter of intense contention and legal wrangling.

Regardless of the games Kenetech and SeaWest played with their supply bids, the end result was promised power prices well below anybody's expectations. In 1994, after letters of bipartisan support from state legislators, and even an endorsement by large industrial customers who typically are not friends of green power, the CPUC finally approved moving forward with the contracts to build the new projects. The CPUC did, however, put the 585 MW bids on hold pending a review of the developers' unusual bidding strategies. Kenetech joined SeaWest in claiming that such retroactive evaluation of bids was improper, and questioned the CPUC's jurisdiction over independent power producers.

Zond and FloWind continued to push for a "real world" evaluation of bids. FloWind was particularly scathing about SeaWest's bids, the most outrageous because they consistently involved negative energy pricing. Describing SeaWest's bids as "behind-closed-door deals which only benefit presumed manipulators at the expense of rate payers and honest bidders," FloWind requested that all of the bids for wind projects be thrown out. SeaWest argued that losing bidders, such as FloWind (which had nothing to show for its efforts) should not be able to challenge winning bids until the CPUC determined FloWind's request for review of the bids was appropriate. FloWind retorted, "It is not hyperbolic to analogize SeaWest's request to that of a bank robber demanding that the bank prove it has suffered harm before the prosecutor can proceed with felony charges." If the CPUC determined that BRPU wind bids must be based on realistic engineering and economic factors, "then 75 percent of the winning bids will be disqualified," FloWind claimed.

"The wind guys are the worst," Smutny-Jones told me during the time of the BRPU bids. "They are a bunch of back-stabbing characters. I've never seen anything like it. They fight amongst each other like cats and dogs. They're crazy."

All of the infighting by wind companies soon became irrelevant. John Bryson, founder of the Natural Resources Defense Council and the former CPUC commissioner who approved the first utility power purchase contracts with Kenetech, took the unusual step of appealing the CPUC decision authorizing the BRPU to the Federal Energy Regulatory Commission (FERC), which rarely got involved with state regulatory issues. Bryson was now head of Southern California Edison (SCE), but had been a state regulator and appeared to be one of California's strongest supporters of wind power. At SCE, Bryson aggressively cultivated the utility's image as a "green" provider of electricity, but he nonetheless succeeded in getting federal regulators to overrule the CPUC and halt development of new wind farms authorized by the BRPU. One argument Edison used was that the imposition of environmental payments was evidence of "systematic favoritism" toward renewable energy sources.

This FERC order did not affect a separate negotiated deal between Edison and Kenetech for 460 MW at even lower prices than the BRPU bids. Clarence Grebe, spokesman for Kenetech added: "There is a fear that absent BRPU-type mandatory purchase programs, wind power loses. From Kenetech's standpoint, the fallacy in that argument is that wind power is not competitive with other resources. We believe we have a technology that is fully competitive with newly built power plants."

Chapter 21

Machines in the Garden

> Current and planned windpower production is not reconcilable with environmental protection and will never co-exist with endangered and protected migratory and resident raptors. Both require exactly the same resource areas and one must be destroyed.
>
> —Raymond Suitor

Americans are full of contradictions. Despite our incredibly wasteful energy habits, Americans consider power plants to be as popular as income taxes. We prefer to compartmentalize. Power stations should be hidden from view, not out in the open on naked hills, many of us believe, if only subconsciously. Wind power exposes this contradiction in the most visible of ways. The whirling motion of wind turbine blades, spinning only when there is sufficient wind, reminds us how much we take our electricity supply for granted. If we relied purely on wind power, as a few of our grandparents did, we couldn't watch TV or listen to the CD player any time we wanted. We would have to organize our lives around the rhythms of nature.

In American culture, reality is driven by technology—the "machine"—while pastoral landscapes and rural dreams persist in our romantic idealizations of the world. At least that's the way Robert Thayer, chairman of the Department of Environmental Design at the University of California–Davis, sees it. Wind plants, because of the way they dominate large, otherwise natural landscapes, expose our notions of undisturbed nature as myth. Their development challenges many notions everyday Americans have about "scenery." The siting of nuclear and fossil fuel plants in isolated locations far from the consumers of the commodity we call "electricity" on the other hand, perpetuates belief in a nature separate and unspoiled by human activity by allowing us to remain unaware of the environmental impact of our electricity guzzling gadgets.

Thayer has conducted extensive research into how citizens respond to California's wind farms. When asked to choose between a wind farm, a facility that burns wood waste, a state-of-the-art natural gas cogeneration plant, or a nuclear reactor, people choose wind. But positive responses can shift when one moves from the abstract concept of a wind farm to seeing the spinning turbines on what was once a virgin ridgeline near one's own home. Thayer's research shows that as people who lived closer and closer to the Altamont Pass were questioned, responses became more and more negative. Thayer sums up the dilemma of the "visual" impact of wind farms in the following way:

> When the large California wind plants were first established, they caused a portion of the public to react negatively, with a significant minority feeling that the turbines represented a violation of the "rural" landscape character and pastoral beauty. But wind energy's visibility can be seen as an advantage if functional transparency is valued. With wind energy plants, "what you see is what you get." When the wind blows, turbines spin, and electricity is generated. When the wind doesn't blow, the turbines are idle. This rather direct expression of functions serves to reinforce wind energy's sense of landscape appropriateness, clarity, and comprehensibility. In the long run, wind energy will contribute highly to a unique sense of place.

Thayer makes an interesting point about new ways to appreciate how wind turbines can be integrated into the gestalt of a particular landscape. His suggestion that these machines imbue a place with unique value is, unfortunately, beyond the grasp of most consumers. But Thayer's research also shows that, in an ideal world, consumers want fewer and larger turbines arranged in orderly and uniform arrays. Since that is the new industry trend, and wind technology has matured from chaotic prototypes to sleek state-of-the-art commercial machines, wind farming's acceptability among the masses may grow.

As any wind farmer will tell you, translating results from the research lab into the field never seems to go as planned. The technology quickly became a lightning rod sparking a sharp debate about different kinds of environmental values. When it comes right down to it, wind farming is all about very deep questions of balance. Balance is important in everything from international diplomacy to our personal lives. The hard part is properly weighing everything. It is particularly difficult when it comes to wind power.

◆

The Columbia River is a magnificent flow that has been the lifeblood of the Pacific Northwest. Dam after colossal dam on this wide waterway, which serves as the border between Washington and Oregon, generates enormous amounts of cheap electricity, yielding to local aluminum factories and other heavy industries located here for the nation's lowest energy prices. The Columbia Gorge also serves as a major transportation corridor. After all, this Mississippi of the West stretches all the way from high up in the Canadian Rockies to the Hanford Reservation, where its waters are used to cool nuclear power plants, before it links up with the Yakima and Snake rivers and meanders further west, spilling into the Pacific.

In its effort to move outside of California and build new markets for its new wind power turbine in the mid-'90s, Kenetech pursued plans to farm the wind in Washington, Montana, and Oregon, places where it had purchased the rights to wind before it even came to California's Altamont Pass. The company chose what it thought was the perfect spot for a wind farm in the Columbia Gorge, a site called the Columbia Hills. It was near land already sacrificed to recreational vehicles, wind surfers, and industry. A favorite tourist destination nearby is a replica of Stonehenge. But a dam, an aluminum factory, and two very large transmission lines all lay near the proposed wind farm region. "The place doesn't look too much different from the Altamont," claimed Peter West, an environmentalist who works for the Portland-based Renewable Northwest Project, whose goal is to educate citizens and policymakers about the environmental and economic benefits of renewable energy. During the summer, the bronze hills do indeed bear some resemblance to the nude ridges of the Altamont.

The initial California Energy Commission report on avian mortality at the Altamont Pass had a major impact on West's work when several new wind farms were proposed in the early 1990s in the Pacific Northwest. The wind power industry seemed to be on the verge of another boom, prompted by the combination of major cost reductions from new, larger wind turbines and the federal production tax credit of 1.5 cents per kWh.

Wind farms were often dismissed by folks in the Pacific Northwest as a California thing, as a technology that had been corrupted by the free enterprise system. West once was called "a lapdog at the seat of greedy industry" when he spoke up for Kenetech's Columbia Hills wind farm proposal. A former colleague of West's in the fight to protect Washington State's most beautiful rivers, also accused him of "perpetuating cultural genocide." The author of that comment, Dennis White, heads up the Columbia Gorge Audubon Society. He led the charge on behalf of environmentalists and Native Americans to stop the Columbia Hills wind farm with this impassioned rhetoric:

> You have chosen to erect your endless strings of steel towers and whirling blades on a landscape where falcons and eagles still soar

> uninhibited above the cliffs and ridges; a place Native people hold in high regard for collection of traditional foods and for calling to the spirits; a place where locals and travelers alike can look out across a beautiful and serene land and ponder innocently its evolutionary making.
>
> This project will kill birds and is being advanced under full knowledge that birds protected under the Migratory Bird Treaty Act, Bald and Golden Eagle Protection Act and the Endangered Species Act will be reduced to shredded corpses. This is a criminal act—no more, no less—and you should be prosecuted now, for premeditatedly violating federal law.
>
> Let me give you a warning: If this project goes forward, our eagles and falcons will die, but we will be there every time, handing the dead birds as evidence to the attorney general, to insure that you will be prosecuted.

West discovered the proposed site of the wind farm was located amid millions of acres of land ceded by the Yakima tribe to the U.S. government in 1858. The tribe is the only party besides Audubon to oppose the wind project, but it opposes all development in the area. West appreciates Native American views on land, time, and appropriate use. Just the same, he notes, the place is "overgrazed as hell." Chief Johnny Jackson, one of four Indian leaders among the River People and the Yakima Indian Nation, claimed a number of sacred sites are located within the proposed wind farm region. "That's where my people go to gather traditional food and medicines in the spring and fall. That's all my people have left. I have lived here all my life, as have my father and mother, grandfather and grandmother. We are not against wind power. This is just not the right place for it."

One of the odder arguments White used in opposing the project is that global warming has yet to be proved—an argument typically associated with advocates for fossil fuels, not those who profess to want to save the earth. "As the timber industry will always want to cut the biggest trees, our old growth forest; as the livestock industry will always want to graze the lushest grass, our riparian areas; and as development interests will always want to develop on level farm land; so too will the wind power industry always want to erect their turbines in the windiest areas. We should expect industry to behave like industry," said White, clearly articulating the suspicion many of those in the environmental community still have for any kind of business development, no matter how "green" it may be.

For the record, Kenetech had divided its 13-mile wind farm land holdings into staged developments. The first proposed stage would have had 85 turbines generating 31 MW of electricity. Each subsequent stage would proceed only if monitoring indicated minimal avian mortality. Over time, Kenetech envisioned the wind farm growing to 345 turbines generating 115 MW of electricity. Federal and state environmental review of the proposed project provided evidence that could be used both to support and to oppose it. A United States Fish and Wildlife Service (FWS) assessment concluded the fully built-out wind farm was "not likely to jeopardize the continued existence of the American peregrine falcon or the bald eagle." It projected a total of 8 to 26 raptor deaths per year from this farm and a neighboring 100 MW wind farm to be developed by Washington municipal utilities. Yet FWS also pointed out that a known bald eagle roost was located within the proposed wind farm area. In addition, a confirmed sighting of two peregrine falcons foraging nearby was significant, since these falcons are a federally listed species. Only seven pairs, all introduced, reside in the entire Columbia Gorge, FWS noted. Two state-listed birds (the long-billed curlew and western bluebird) were also spotted on the site.

Still, the Kenetech site was determined to be "not an important bird area" by wind farm proponents. According to Hawkwatch International, the average number of sightings for "good" hawk viewing areas in western North America is 10.5 per hour; this Columbia Hills site measured 3.6 hawks per hour. "There are no renewable energy projects in the Pacific Northwest, except very old hydroelectric power plants built over 50 years ago, and a few biomass facilities," said Rachel Shimshak, director of the Renewable Northwest Project. For this reason, she found it ironic that local environmentalists were "setting up a standard far more stringent for wind power than those that apply to natural gas and coal plants." She continued, "Opponents are not looking at the bigger picture. Here's a 31 MW wind project and you go just up the river and there are three natural gas plants representing over 1,500 MW that have been installed within a four year period." No one seems too concerned about the negative environmental impacts of an overreliance on gas, she complained. "When they drill for gas roads get built and these roads make it easier for loggers to come in. Sure, gas is cleaner than coal, but it still releases a lot of carbon dioxide," she said.

The three most recent gas plants alone, located in Vancouver, Washington, and in Hermiston and Boardman, Oregon, will emit 3 million tons of carbon dioxide, 31,500 tons of methane, and 450 tons of nitrogen oxides each year of operation, she said. In contrast, a 115 MW wind farm would displace annual emissions of 270,000 tons of carbon dioxide, 2,240 tons of sulfur dioxide, 1,150 tons of methane, and 1,060 tons of nitrogen oxide. West added, "Local environmentalists wanted to make this wind project a model of perfection. They've already got 80 to 90 percent of they wanted. Now they are asking for 95 percent. The remaining 5 percent could kill the project!"

West's prediction came true. Though most of the Kenetech site, excluding the one section identified as having sensitive bird populations, was ultimately purchased by Zond, there are no current plans to proceed in building a new wind farm.

◆

To many observers, Rattlesnake Ridge is hardly a landmark worth capturing on film or even acknowledging in one's nature journal. Set in the brush and wheat fields of eastern Washington State, the dry rocky ridge has none of the lush forests and streams that are hallmarks of the glorious Pacific Northwest typically portrayed in postcards.

Nevertheless, when Kenetech proposed a project at the ridge, local citizens organized in defense of the ridges. Led by Clifford Groff, a councilman from the town of Kennewick, they formed Save Our Rattlesnake Environment (SORE). SORE enlisted help from the Sierra Club and state wildlife officials to kill the proposal.

The ridge, the tallest geologic feature in the region, is "ruggedly ugly," said Groff. It deserves protection because it "has personality." The Washington State Department of Fish and Wildlife offered a more concrete assessment of the ridge. It lies within the largest nature research area in the state, several birds considered threatened or endangered can be found there, and less than 10 percent of the native arid landscape it typifies is left in Washington.

Groff works at the Hanford Reservation, a 570-square-mile complex of rusting, radioactive monuments to America's legacy of nuclear weaponry. This is, after all, the site of the reactors that produced the plutonium for the bomb that destroyed Nagasaki, and today is the site of nearly half the country's defunct nuclear arsenal. "We have a poor old nuke just sitting there, ready to produce 1,250 MW of power, and here was Kenetech/Windpower talking about destroying a unique ecosystem to produce only 100 MW for people living in Idaho, Oregon and Seattle," Groff complained. Rattlesnake Ridge is actually located within the boundaries of the huge Hanford Reservation, where siting something that generates power has never been much of a controversy before. Kenetech had its reservations about siting a wind farm so near to the thousands of rusting, corroding barrels of radioactive waste in storage nearby. But in Groff's view, "there is no problem with the waste." He points to France as a country that receives 80 percent of its power from nuclear plants. "That's what we should be doing."

William Whalen, head of the National Park Service under President Jimmy Carter, was a Kenetech vice president at the time. His broad, square face breaks into creases of discomfort at the mention of Groff. "These people pop up everywhere we go," said Whalen. "They are zealots. Whatever their motivations, they get up in the morning and the first thing they are thinking about, even before that first cup of coffee, is stopping wind power." And while locals complained about the dust that might get kicked up by the wind farms on Rattlesnake Ridge, Whalen worried about the thousands of barrels of radioactive waste that surrounded the proposed site. "They told me the barrels burp every once in a while, and these burps could be a significant event," related Whalen. "And they were worried about a little dust," he continued, shaking his head in disbelief.

◆

"This would be one way of destroying the last best place. We would effectively destroy the eagle population in the area." These are the words of Ron Wiggins, a resident of Big Timber, Montana. He's talking about Kenetech's plans for harvesting wind in Big Sky Country, the State of Montana, which prides itself on "open space and peace of mind." With a background in aerodynamics, Wiggins once liked the idea of wind power. However, he always viewed it as a small-scale technology, a view still shared by other well-meaning folks. When he heard about plans for hundreds of large machines in his admittedly large backyard, he started to investigate. He didn't like what he found and he broadcast his message to the local and the national press. "People didn't want to see hundreds of eagles get slaughtered," he stated, referring to a proposed 150 MW facility earmarked for Livingston.

Local residents were also concerned about impacts on property values, Wiggins said. Interesting that he should bring up real estate concerns. Wiggins works for Raymond Suitor, a real estate developer who serves as the U.S. representative of Country Guardian, an anti-wind propaganda group formed in the fall of 1993 and based in the United Kingdom. One of the group's original chief officers is Sir Bernard Ingham, former press secretary for Margaret Thatcher and former PR man for British Nuclear Fuels. Bernard does not hide his allegiance, authoring stories with titles such as "Nuclear power is greener than windfarms." Wiggins claims he speaks for Suitor, and he was not bashful when it came to his energy vote. "The public is nuts to put the ax to nuclear power," Wiggins offered when asked about alternatives to wind power. "Nukes are the only way to go if you handle it right."

"When you consider the wildlife and visual impacts, it's every bit as bad as the coal-fired plant," Wiggins said of wind power. Indeed, he argues that a coal plant may be better for the environment. He told me a local coal plant uses up 30 square miles of land—for everything: power generation, mining, reclamation—everything. To produce the same amount of electricity from wind farms, it would take 500 square miles of land. "Wind power is a bad idea. It is just so inefficient. The only reason we are building any of these wind farms is that they are politically correct," he said. "I have a family and five kids. And I have to manage this place. Meanwhile, the bureaucrats pushing this wind power stuff have all the time in the world. We are at a disadvantage because we don't know what goes on behind the scenes," he complained. He freely acknowledged that he has kept in contact with wind power opponents from Maine, California, Washington State—indeed around the world.

What started Wiggins in his personal crusade to stop wind farms was the purchase by Kenetech of an easement on a 5,000 acre property adjacent to Sweet Grass Farms, a 1,400 acre ranch that grows specialty crops, like gourmet potatoes. Suitor was thinking of opening up a private hunting and fishing club. The thought of having strings of wind turbines nearby gave him, and Wiggins, a bad case of indigestion.

Americans like Suitor, who once also wanted to develop 50 acre ranchettes near the site of the proposed wind farms, have been used in the fight to slow wind power overseas, where government policies limiting carbon dioxide emissions linked to the threat of global warming have thrust wind power into the limelight. "Wind-generated power stations head the list of 'dirty' dangerous and destructive form [sic] of energy production," reads one letter Suitor sent to British government officials. "I trust British citizens will awaken and stop this ugly, destructive blight. Windpower is NOT GREEN! It is black, greasy, deadly and destructive."

Despite the efforts of Country Guardian, wind power development in Europe has continued to spread at a far more rapid pace than in the United States. One reason is that European wind farms tend to be smaller than projects proposed in the United States, and governments there have far stronger targets for meeting carbon dioxide reductions; the United States has yet to set any firm targets. Because many European countries have done so, new wind farms, the most cost-effective of today's renewable bulk power options, have a high priority. In addition, the wind farmers in Europe are increasingly siting extremely large wind turbines offshore, in bays and harbors, where environmental concerns have been minimal.

◆

Though a wind farm was eventually installed in eastern Oregon with the support of both the Audubon Society and the Confederated Tribes of Umatillas, this project was the exception rather than the rule in the Pacific Northwest. Conflicts among environmentalists over wind power became more frequent as its appeal spread outside of California. The controversy over bird kills continued to plague an industry already crippled by the double whammy of low fossil fuel prices and uncertainty about new proposals to deregulate the power market. Many utilities began to reconsider wind farms they had just recently viewed quite favorably. Ironically enough, the low prices offered by wind and other bidders in the BURP helped build momentum in California to do away with the electric utility monopolies that had been in place since just after the turn of the century. One of the goals of the restructuring process was to get utilities out of the generation business, a trend that threw yet another kink into Kenetech's long-term business strategy of selling turbines to utility buyers rather than installing and operating them as developer/owners.

Yet the concerns over protected birds, particularly raptors such as eagles, hawks, and condors, was the most immediate threat to a Kenetech now trying to deliver on the promises made to its new public shareholders. Dale Osborn confessed that the magnitude of the bird issue caught Kenetech off guard. "Cynthia Struzik was more damaging to Kenetech's plans than anybody else. She brought many of the projects we were counting on to a screeching halt. I mean, she sent out a letter to the heads of utilities considering wind projects claiming they could end up in prison if they authorized construction of a wind farm that killed a single bird!" Osborn noted that the Struzik letter convinced Public Service of Colorado to pull out of a proposed Wyoming project. "Her letter slowed everything down for one full year," said Osborn, referring to the variety of projects Kenetech had proposed throughout the West.

Kenetech did succeed in building the first wind farm in Minnesota. A last-minute legislative deal in 1994 required Northern States Power (NSP) to add 425 MW of wind power to its supply mix. In return, NSP was allowed to store spent nuclear fuel in steel casks outside the Prairie Island nuclear plant. Another 400 MW of wind power was also on the table if state regulators determined that wind was the lowest-cost electricity-generating technology once environmental costs were factored into the life cycle cost.

The Midwest was beginning to shape up as a major wind power market. Kenetech wanted to make certain it got most of the action in the largest utility power purchase opportunity in the country. It quickly bought the wind rights to land that looked like the best wind farm sites and started calculating the turbine sales revenue these projects would generate. Kenetech looked at Buffalo Ridge, which on a map looked like a crescent moon in the southwestern corner of Minnesota, as its next Altamont Pass. The first project was only 25 MW and Kenetech won the bid. It was to be the company's first commercial installation of its 33M-VS turbines in the Midwest.

One bright spot in the Minnesota story was that the bird issue was being addressed right up front. A state known for its environmental ethic, Minnesota pledged to develop the most comprehensive avian monitoring program in the nation to insure new wind farms did not get placed in migratory paths or near habitats of protected species. "The environmental community of Minnesota is supportive of renewable technologies and especially enthusiastic about wind energy," reported Michael Noble, executive director of Minnesotans for an Energy Efficient Economy, one of the key players that helped negotiate the legislation creating the wind power mandate for NSP. Even the Audubon Society supported wind power development along the entire Buffalo Ridge region, the best wind farming sites in the state. So far, more bats have been killed than birds, and neither bats nor bird populations seem endangered by the new wind farms.

Kenetech also received help from the Audubon Society in Texas, another new prime market for wind power. "Kenetech did Texas right," said Tom "Smitty" Smith, executive director of Public Citizen, noting that Kenetech went into the local community to find an individual "who was a long-term environmentalist who was able to bring a large number of folks together." Dede Armantrout, regional vice president of the Audubon Society, was impressed. Kenetech allowed local environmentalists to choose a biologist they trusted to screen areas that raptors and other birds of concern might frequent. "We know the person we chose is scrupulous and detail-oriented—he won't hedge the data," acknowledged Armantrout. "Kenetech has been really open. They seem to be endorsing a solution-oriented approach. I am optimistic and hopeful that we can bring this clean alternative to other forms of energy to Texas."

Chapter 22

Fall from Grace

> The income tax has made more liars of the American people than golf has.
>
> —Will Rogers

The next phase of the Northern States Power wind program in Minnesota was a high stakes game for America's two leading wind power companies. This phase consisted of a large, 100 MW wind farm, up for competitive bid. It was the largest solicitation earmarked for wind power ever in the Unites States. Most in the wind industry thought Kenetech the clear favorite to win the right to build this second phase, as well as the third and any future phases. Kenetech thought so too.

But it was not to be. Zond won the second phase with a stunningly low bid of 3 cents per kWh!

Zond's bid, of course, included a 1.5 cent per kWh tax credit; the real price was 4.5 cents per kWh. Nonetheless, the American Wind Energy Association and a chorus of renewable energy advocates trumpeted the bid as a sign that wind was getting within the range of being competitive with fossil fuels. Kenetech had focused so much of its marketing on the five-cent machine, but fossil fuel prices, particularly for natural gas, continued to decline. Flooding the power supply market was a new generation of super efficient gas-fired turbines that could produce electricity in the 3 cent per kWh range. The promise of a five-cent machine had been obviated.

In July 1995, after the Zond award, Kenetech's vice president Michael Alvarez fired off a caustic letter to the Minnesota Public Utilities Commission (MPUC) to cry foul. A sure contract it seemed had been snatched away by its chief rival. Not only did Zond bid a project with a new American wind turbine that no one had seen, but it proposed to build the new wind farm on land whose wind rights were already in Kenetech's hands. Zond proposed to have the state condemn the property, a proposition that infuriated Kenetech but that state regulators ruled was entirely proper. Kenetech protested, "There is no legal precedent whatsoever for the condemnation of wind rights owned by one private party for the benefit of a competing private party with whom the condemnor wishes to do business." Alvarez went on to charge that the selection of Zond "endangers the credibility of wind generation as a generally accepted utility technology. As the historical record amply reflects, the wind power industry has been replete with exaggerations of capabilities and remarkably short on performance."

Zond had built its business by relying on Vestas turbines, one of the top producing Danish machines. But it had changed its business strategy in the early 1990s and began to develop its own wind turbine. Although relying on the same basic three-bladed, upwind, active-yaw design as did the Danes and now Kenetech, Zond's so-called Z Series turbines did employ a few unique features. Instead of pitching the entire blade, the Zond turbine relied on ailerons on the blades to control power output during high winds. It also had an integrated drivetrain, which was touted as a way to cut manufacturing costs. The Z-40 had a 500 kW generator and reached that power level when winds reached 27 mph. The main advantage the Zond turbine had over the Kenetech machine was that its rotor swept a 47 percent larger area, which translated into potentially much higher power production. The turbine Zond proposed for use on the NSP project—the Z-46—was even larger than the Z-40. However, neither the Z-40 nor Z-46 included the variable-speed innovation that was the distinguishing characteristic of Kenetech's 33M-VS.

Kenetech was not alone in wondering about the credibility of Zond's bargain basement price for wind power. Even those who understood that the three-cent figure incorporated the federal production tax credit of 1.5 cents per kWh still shook their heads skeptically. "No one has yet proven that they can successfully build and operate a 5 cent per kWh windmill for any extended period. . . . Will we soon be promising that wind energy will be too cheap to meter?" asked Paul Gipe, a long-time industry insider. Here's how he put the Zond offer in perspective:

> For the record, Zond bid the contract with the Z-46. This is a windmill that does not now exist. It is a paper wind turbine. Zond has never built a Z-46. Moreover, Zond has only built two (yes, only 2) prototypes of its Z-40. Thus this contract is dependent on the significant upgrading of a wind turbine that has yet to prove itself in the field.
>
> There is only one European wind turbine company currently

> manufacturing commercial wind turbines greater than 600 kW. And they have only built a few of those. In the NSP contract, we have a new manufacturer proposing to build and install 143 wind turbines by the end of 1997. For perspective, the 100 MW NSP contract is more than the annual installations of Denmark and the Netherlands combined and nearly as much as installed in all of the UK. This is a big contract and a highly public one. Should the project fail to meet NSP's expectations in any way, it will severely damage the wind industry's hard-won credibility in the US.

Perhaps the deepest irony in Kenetech's letter protesting the Zond award is that Kenetech accused Zond of the very tactics that many attributed to Kenetech: wild promises of reduced costs, poor-performing prototypes, and "paper" turbines.

A report filed by Henry Hermann, an investment advisor, the previous summer (in August 1994) underscored the surreal nature of Alvarez's comments, and the omens on the horizon that might have added to them their urgency. Hermann had paid a visit to the Kenetech site in Minnesota where Kenetech had installed its 33M-VS turbine. What he witnessed convinced him to issue a "sell quick" recommendation on the company's stock. "What I saw and heard was in great contrast to the company's public pronouncements. Expectations regarding the company's future may be far too optimistic, and I believe that disappointment could cause a significant decline in the stock price." While Kenetech had reported that the wind turbines selling electricity to NSP were available to generate power between 90 and 100 percent of the time, Hermann saw something else. Sixteen of the 73 wind turbines were not spinning on a day when winds were blowing between 12 and 15 mph. This meant what industry folks call the "availability factor" was only 78 percent. He went on, "While standing on the road by some of the operating wind turbines, the loud groaning, clanging, whining noises emitting from several of the machines strongly suggested to my untrained, unscientific ear that meaningful problems may exist with some of these machines." Far more disturbing, Hermann said, were the three turbines that were missing blades. There were also blades on the ground "with clearly visible cracks, where the blade joins with the hub." But Merrill Lynch quickly responded, claiming that reports of Kenetech's demise were "greatly exaggerated." Problems with blades and hydraulic systems were common start-up problems for any new technology. Merrill Lynch stood firm on its positive stock outlook.

Kenetech's new and improved turbines looked great on paper. But, then, so did the mammoth machines funded years earlier by Divone at DOE. Despite its sophisticated digital electronics and computer software requiring almost 150 individual patents, Kenetech's variable-speed rotor wasn't delivering on its purported 10 percent increase in electricity production. Furthermore, the new level of complexity introduced by the digital and computer features just created new technological challenges that weren't quite resolved. Efforts to cut costs apparently backfired, too.

In fact, the 33M-VS, on which Kenetech had staked so much, was about to become the final straw.

Like the vultures that also circle the Altamont, attorneys for the thousands of investors now owning Kenetech stock, alerted by Hermann's report and 1993's less than spectacular business results, had begun to sort through its complex web of transactions. Tinkerman's claims about Kenetech's numbers games suddenly found their ear as attorneys representing aggrieved investors pursued a securities fraud case claiming, among other things, that the firm's last few public offerings (the final one, of preferred stock, ending in August 1995) inadequately disclosed technical problems with its highly touted new turbine. Implicated in the investors' class-action lawsuit were not only individuals such as Stanley Charren, who stepped down as chairman of the board in 1995, but Merrill Lynch and the other investment bankers that had helped make Kenetech such a hit among private investors.

By March 1996 the value of Kenetech's stock had declined from highs of over $29 per share for common and $20 for preferred to just one dollar for both. All told, investors had purchased some 36 million shares of common stock and over 5 million of preferred. The total drop in value was cataclysmic, amounting to more than $1.1 billion. Both the company and the underwriters were named as defendants, on grounds that they had artificially inflated the value of the Kenetech securities in a coordinated campaign of misinformation amounting to fraud. Merrill Lynch and its investment banking compatriots were accused in the lawsuit of knowing Kenetech's problems as the offer of preferred stock was being issued. They were alleged to have remained mum because of the lucrative fees they enjoyed from these large financial transactions. Merrill Lynch and companies like Smith Barney and Morgan Stanley (eventually more than 30 firms were involved) had sold $92 million worth of notes in December 1992, netting banking fees of at least $2.75 million.

By some accounts, Kenetech was simply very good at manipulating perceptions among the underwriters. The picture painted in the lawsuit was that the firm went so far as to hold regular conference calls to disseminate false information to investment analysts, who then became conduits of misinformation that in turn encouraged positive press and kept Kenetech's stock price high. But the class action suit alleges numerous violations of the federal Securities and Exchange Commission codes and claims that even the underwriters soon learned some of the truth about Kenetech's growing woes. For example, as early as February 1994, a conference call was held among Kenetech, Merrill Lynch, and other underwriters. It was allegedly revealed that the marketing plan for the 33M-VS was just getting underway, but due to numerous unresolved problems, the turbines would cost at least 20 percent more than originally projected. Coupled with the vagaries of the wind and the tightness of the power market, this meant that Kenetech really had no ability to predict future net revenues accurately. Despite these disturbing details, many in the large retinue of underwriters were participating in road shows trumpeting Kenetech stocks.

Also according to the claim, suppliers of key turbine parts warned Kenetech that their warranties would no longer hold if Kenetech operated the machines in winds greater than 44 mph. Moreover, the purported benefits of variable-speed technology had been overstated. The infrequent times that winds did exceed 44 mph, most of the increased energy captured by the variable-speed rotor was apparently lost during the energy-intensive process of conversion into steady-frequency conditioned utility power.

And last, but certainly not least, were the bird issues Osborn noted earlier. The lawsuit specifically referenced these problems:

> At the time of the IPO Prospectus, defendants knew, or recklessly disregarded the fact, that existing Kenetech wind turbines were killing numerous eagles and other migratory birds protected by federal statutes. Nowhere in the IPO Prospectus did defendants reveal that avian deaths and concern over such deaths had prevented development of a major wind site by Kenetech in Montana and that Kenetech was experiencing serious problems because of the avian mortality in development of its wind site in Tarifa, Spain. In addition, defendants knew at the time of the IPO that such avian problems would constrain Kenetech from replacing 56-100 turbines with 33M-VS turbines in the Altamont Pass and would have a serious impact on the development of a proposed windplant in the Columbia River Gorge (Columbia Hills). Such environmental problems were also having a significant impact on Kenetech's ability to develop new windplant sites in the United States and internationally.

As early as August 1993 the firm had realized its projections for growth issued in the previous October overstated sales by 35 percent. Instead of a $4 million profit in 1993 the company now was looking at a $5 million loss. Part of the problem was not just the new turbine, but the old ones. Not only was the information Kenetech was releasing about the 33M-VS suspect, but the maintenance and operation costs for its 56-100s were consistently higher than projected, and, like Fayette, much of the money raised ostensibly to build the future was apparently being used to fix past mistakes.

This last point, of course, was no surprise to Tinkerman, who claimed all along that Kenetech never possessed a wind turbine that worked as well as portrayed by the company. "The very first installations of the 56-50s resulted in 100 derelict machines that were destroying themselves on an hourly basis. I know. I saw it with my own eyes," Tinkerman once told me. I asked him: How could Kenetech survive with such shoddy machinery? "This is the inside story," he said, motioning with his hand at the smoke from his faltering cigar. "They erected a cluster of their new and improved 56-100 wind turbines in a place called 'screamer alley,' one of the best wind sites in the Altamont Pass. Internally they projected these turbines would generate as much as 300,000 kWh per year. The first year's performance was 214,000 kWh—more energy than they had projected for the average Altamont sites: 210,000 kWh. What Kenetech neglected to say was that 1984 was an incredible wind year. These turbines, given their specific location and the quality of the wind fuel in this particular year, only produced two-thirds of what they hoped for. All of the 56-100s installed between 1985 and 1988 were sold by Wind Potato on the basis of this 1984 'screamer alley' performance data. They sold projects on the average, while reporting as average the screamer sites."

Tinkerman then hit on a larger point: "They suffered from a symptom of American business: the bean counters always overrule the engineers. That is the best case. The worst case is fraud, products, such as wind turbines, hyped well beyond any sense of reality."

But he also suggested that, "If Kenetech could have postponed the release of their 33M-VS for one year, and made changes based upon their engineering data, *and* stopped using new offerings to pay off the old debts from projects that never performed, they would have never gone bankrupt."

◆

One of the more objective assessments of the demise of Kenetech comes from Ed DeMeo, a long-time director of R&D at the Electric Power Research Institute (EPRI), the research arm of the United States electric utilities industry, which designed the computer electronics in the 33M-VS. DeMeo recognized there were forces beyond the company's control, concerns over avian mortality being one. Another was the restructuring of the electricity industry, which caused Southern California Edison to kill what were to be the first wind projects to be built in California in the mid-'90s. DeMeo has a much longer list, however, of forces *within* Kenetech's control. The biggest problem was the corporate culture and management style, which DeMeo described as "externally and internally arrogant," a defect "which neutralized the many difficult things the company did well." There was little trust within the company and within the larger world of wind farming professionals. "The lack of a team atmosphere crippled a very talented staff," said DeMeo, noting that the egos of top management drove talented people away and led to decisions that went against sound technical judgment. "They had very few friends. Many of the ex-employees wanted wind power to succeed but Kenetech to fail." Kenetech even alienated EPRI, which had helped get the 33M-VS turbine development program off the ground. The dispute revolved around Kenetech's payment of royalties to EPRI for the aforementioned electronics.

According to DeMeo, the final factor in the fall was that upper management badly misread how power markets were evolving. They authorized maintaining a large and costly marketing campaign directed at utilities well after the onset of electric utility restructuring, which should have signaled a shift in the target market from wholesale utilities back to the retail customers.

Oddly, Osborn echoes the criticism that upper management lacked appreciation of the people who could come up with sound and proper designs. "Our CEO [Alderson] viewed engineering as a necessary evil. He never appreciated that the core of a competitive advantage was technical staff. The culture at Kenetech was to keep the warfare going between the engineers and the finance types. The single most fundamental factor [in Kenetech's fall] was building up $150 million in inventory of wind turbines for projects that were not yet permitted," said Osborn. He was referring to the wind farm proposals delayed or killed in the Pacific Northwest because of concerns about birds, but which Kenetech continued to count on. Furthermore, those machines sitting in inventory still had a few bugs to be worked out. From Osborn's vantage point, it would have been better to keep inventory down, since the machine was still being perfected in the field. "The reason why Dell computers is so successful is that they have no inventory," he remarked.

Osborn still defends the basic design. "There were no fatal flaws," he insisted. "Engineers are never done. They always know that they can make it a little better. The technology did not fail. We produced a five-cent machine. The key problem, in retrospect, was the poor quality of senior management decisions. But other than Charren, the board of directors never understood the complexity of wind technology. The board had no marketing or engineering expertise—it was not diversified." A pang of remorse came through from Osborn at the conclusion of our interview. "What is most troublesome was the fate of the manufacturing folks. They put everything they had into the company stock. Lives were destroyed with the fall of Kenetech. You know, the people who worked here didn't work for Kenetech. They worked for the promise of a legacy of a better environment, for stewardship of our natural resources." He gazed out of the window before saying one last thing. "The bottom line on Kenetech was that management believed their own B.S."

Stoddard had a different story. "They could have taken my design—which still works—and made it their product and it would have worked just fine. But they had to tinker and then took some Draconian steps. And when they couldn't put off introducing a new product, they released it and it was a disaster." Part of the problem at Kenetech was its pervasive cult of secrecy. "It was difficult to get information from even the other side of the shop. Management was obsessed with the idea that other companies would steal their ideas. And then there is the 33M-VS, which was designed by an in-house committee. The hubris was palpable. They thought they were the only game in town. No one else counted. It was Osborn who ordered Kenetech employees to reduce the size of the blade roots. The blade failures led to pitch link problems. It was like dominoes. . . ." There was a long silence.

As CEO, Gerald Alderson probably tops most people's lists when it comes to a talent for finger pointing. Alderson's view of the Kenetech collapse reflects his fundamental free enterprise values—as well as his disdain for engineers. "The reason we went under is very simple. There was a latent engineering defect—well, actually more than one—in our new wind turbine. Some will argue the cause of our demise was the deregulation of the electricity market, but I'm telling you that just made the future market for wind power less robust. That's looking at the upside. From my vantage point, the bankruptcy was 100 percent the engineering problems we had with the 33M-VS," said Alderson, who now runs a company that publishes a national yellow pages directory of electricity suppliers, as well as another firm looking to become a national retailer of electricity services, including green power.

Jamie Chapman is a highly respected engineering consultant who helped transform Woody Stoddard's work on wind turbines into the machine that powered Kenetech's rise—the 56-100 kW turbine. Chapman does not fault Kenetech's decision to switch designs. From his vantage point, as former vice president of operations and maintenance, the evolution of Kenetech's turbine design was a very logical progression. "In the beginning, we didn't have the knowledge about the gust structure of the wind and how that impacted braking materials and the fatigue lifetime of wind turbines. We didn't have the insights within such a short time scale, or the same quality modeling tools, as today. What the Altamont Pass and California's other wind resource areas were was a very large laboratory of prototype machines."

Chapman summed up the rationale for the radical shift in design: "We saw the path of the future was larger wind turbines. For technical reasons, we switched to an upwind design. Power quality was a major concern of utilities and we thought a variable-speed wind turbine could better control torque spikes from gusts and therefore be better able to address concerns regarding wind power's ability to generate utility-grade power. We also saw the cost of power electronics dropping rapidly over the next five to ten years, which would also allow for future cost savings." He acknowledged that the machine may not have been ready for prime time. "It would have been desirable to test five to ten machines for a year before widespread deployment. Nature will always tell you what you overlooked." Chapman resigned from Kenetech, however, just as the effort to sell the 33M-VS was picking up steam. He has consistently refused to say why he left.

Fayette's John Eckland can empathize with the need to install new wind turbines in the field before they are completely ready for the market. "Wind power companies always have enormous cash flow problems and the engineers always ask for another six months. Well, in another six months we'll all be dead. If you listen to the engineers, you'll never be ready," Eckland argued, adding that only through actual field experience can sufficient progress be achieved in perfecting wind power technology.

Charren, the godfather of Kenetech, and described by Stoddard as "the Howard Hughes of the wind business," agrees with Eckland—more or less. When I brought up Eckland's comment about engineers and delays, he started nodding his head. "Let me let you in on a little secret," he said, as if talking to a son, reminding me a bit of Groucho Marx, "Bringing technology to the marketplace is an iterative process. Anything you build now will be built better. They will no longer have to start from scratch. It's really that simple. I don't care if you are talking about an airplane, refrigerator, bike—or a windmill. The reason these other products work so well is they were built over a long period of time. The products create a culture and within this culture is where the learning, the innovation happens, within this much larger process. It is interrupted periodically by an individual or group willing to take risks with a particular product.

"So to get back to your question, this is what I have to say. Anyone who says to you the product isn't ready doesn't know what they are talking about. When are we ready to market the product? Now, that's a different question," he said. "Windmills have been used since the 1600s. How difficult can they be to build? Well, none of the large DOE machines worked. And who built these wind turbines? These were competent companies, some of the biggest names in U.S. business. The problem with wind power is that the math is insidious." He paused. "If you do the math on a clean sheet of paper, the math tells you to build the turbine as big as fucking possible. The problem is, forces at the blade roots are extremely high. The design team pushes and pushes for larger equipment because the math is insidious, very insidious. . . ." He trailed off. The reason why the company had been so successful was that "we did everything. We never sold our wind turbines to third parties. You learn the most, and the fastest, in a wind farm where you have 100 turbines that you build, operate, and take care of. It is a closed loop learning process."

Charren agreed with Chapman that the new, bigger wind turbine "was on the same design path" as the original turbine. "In fact, the very first wind turbine design for USW had a variable-speed component." What led to the downfall of Kenetech, Charren claimed, was that the company was suddenly growing so fast that new sets of problems arose with a frequency that disrupted the learning curve that had served the company so well, and compromised the company's "capacity to answer to new challenges." "When you start putting projects in

Costa Rica, people don't speak English. It's not the same as the Altamont, where the operations and maintenance crew talk every day at 4:30 about what they learned about the machines in the field. We grew into an international giant, but we were still a relatively small company. Then the rules changed with deregulation and the market changed and the technology was more sophisticated, too. We just didn't react in the best way to all of those simultaneous changes.

"In a perfect world, everything would be perfect and Kenetech would still be in business. Well, the world is not quite perfect," said Charren. "The measure of any corporate organism is its ability to grow, mature, and prosper. The key is how the system responds to change." Trying to sound upbeat, he added: "The people who used to be in Kenetech are still in the wind business. They responded. They are still alive. Large companies are collecting small companies—it is a cycle of change. Every problem is a group of problems—and not only do you have to respond, but at the right speed."

Charren left Kenetech in 1995, not because he saw the company going down the tubes but because of his age—he was 72. "I've made a lot more money in other industries, but the wind company was always my most important company," he lamented. His face suddenly took on a pained expression when I mentioned the investor lawsuit that Tinkerman was so diligently assisting. "I take great exception to the charge that I screwed investors. I am very offended," he said.

Chapter 23

Changing of the Guard

> A fossil-fuel civilization is a dinosaur devouring its own tail: it will eat itself to death; the only question is when.
>
> —Daniel Berman and John T. O'Conner

Kenetech's fall was interpreted by many outside observers who had long been skeptical of wind farming as the nail in the coffin of the U.S. wind industry. Kenetech had performed such a high-profile sales job, capitalizing on its spectacular advances, that taking wind power seriously now, after the allegations of fraud, was a leap of faith only a shrinking number of true believers would take. If the company with backing from the largest utility and the financial muscle of Merrill Lynch couldn't make a go at farming wind, then what were the chances of the smaller ventures?

Those within the wind industry simultaneously applauded and groaned when the news of the Kenetech bankruptcy hit. On the one hand, many felt Kenetech got what it deserved; the firm's cavalier attitude had long ago alienated its rivals. On the other hand, the Kenetech bankruptcy made everyone else's job in the wind industry that much harder.

Whereas Kenetech used a megaphone to tell everybody about its wind turbine and its grand plans to dominate future power markets, Zond Systems was always a company leery of publicity and largely unknown to the public at large. By default, Zond was now America's great hope. SeaWest was still around, but it seemed to be more of a financial clearinghouse for wind projects than anything else; it used anybody's turbine and upgraded and refurbished neglected wind farms. With all the bankruptcies, there was plenty of this kind of work around. Zond, however, was following in the footsteps of Kenetech by promoting its own new turbine. Zond also purchased the Kenetech variable-speed patent, and was planning to integrate some discrete aspects of power electronics into its new machines. Of course, because of the way Kenetech had gained its patent, Zond was now able to keep Enercon and any foreign variable-speed wind turbines out of U.S. markets.

After the termination of the Duskin and Cashman investment tax credits, in 1985 and 1986, respectively, Zond, too, had become concerned about its future. "We had a splendid year in 1985. We installed 1,100 turbines in four different locations, took in $170 million and turned a $27 million profit," James Dehlsen recalled. "Our strategy throughout the tax credit years was to show enough of a track record to issue an IPO. That was still our strategy in the beginning of 1986, but our underwriter—Dean Witter—was slow in getting that effort off the ground. By the middle of that year, oil prices had fallen from $30 per barrel to the low teens. That put a chill on the market and we had to withdraw the IPO."

Without the IPO, which was meant to raise $25 million, Zond was in sorry shape. "The only thing we could do was batten down all of the hatches. We just focused on making everything more efficient. By relocating turbines, increasing rotor size, and other upgrades and maintenance techniques, we tried to glean every last kilowatt hour we could out of each turbine," said Dehlsen. The effort boosted performance of Zond's existing wind farms, which now surpassed the 200 MW mark by 20 percent. Zond did have 450 MW in SO 4 contracts, whose energy prices were still very attractive, since the dock had not started yet and they were locked into John Eckland's erroneous oil price forecasts. "The best-looking project was Gorman," Dehlsen said, "and I spent one whole year trying to make Gorman happen." This effort turned into a nightmare when the project was killed for fear of potential California condor casualties. "Once that fell through, there was the Sky River project. Because of its remoteness and the extremely rugged terrain, everybody told us to forget about it," Dehlsen continued, noting that it was one of several potential wind power sites Zond had purchased the rights to in the Tehachapi area. Zond could not heed this well-meaning advice because the project had become "do or die." "We were dealing with the very real prospect of not being able to pay our debts." Dehlsen confirmed that the day after the Gorman project was unanimously voted down he talked to new Kenetech president Osborn about selling the Zond assets. He's now glad he thought the better of it.

While in the early stages of planning for Sky River, Zond discovered that Southern California Edison had already used up all of the space on nearby transmission lines, space Zond had been counting on to ship its wind power into the grid. This meant that even if Zond were able to install the 77 MW project, it had no way of delivering the electricity generated to a utility end-user that had no financial incentive to interconnect with the wind farm. Ultimately, Zond teamed up with SeaWest in a joint venture that cost the two companies $30 million to build a 75-mile, 220 kV transmission line through incredibly harsh landscapes to hook the turbines into Edison's grid. This was the first time private independent power developers had built their own AC transmission lines in order to make a power sale. There was another major milestone for this challenging project: It was the first wind farm that institutional investors sank their money—$160 million—into without the lure of tax credits. One reason they did so was that Zond had advanced the art of measuring and understanding wind fuel. "In the early '80s, there were wildly optimistic projections on power production, perhaps to entice investors. We at Zond tried to use our best judgment. However, instead of installing just a few anemometers for each wind farm, at Sky River we placed one for every three to four turbines," Dehlsen said.

◆

After Sky River was in the ground, Zond was finally on stable footing. "We could have clipped coupons and survived on a secure but modest revenue stream. But we decided to do something more," said Dehlsen. "It was at this point that we decided in order to compete globally, as well as domestically, we had to cut out the profit margin that was now going to the turbine manufacturer. We had to become a vertically integrated company."

It took an intense focus on sound economics to steer Zond through this transition during a time when wind companies were dropping left and right. There were economic advantages to becoming vertically integrated. That was, after all, why Kenetech kept going for so long. But one could also make the argument that such a strategy made more sense in the early '80s, when turbine designs were in an immature state and learn-as-you-go was the order of the day. Now, in the late '80s, there were several turbines on the market that already worked quite well. Virtually all of Zond's projects relied upon Vestas, the Danish company that remains the largest turbine manufacturer in the world to this day. The superior performance of Vestas machines, together with the fact that Zond retained financial interests in all of its wind farms, provided a long-term source of revenue for the company's core operations. Kenetech never benefited from this kind of stream, since more often than not it did not own the projects it put in the ground, and the operation and maintenance costs of the farms it did own often exceeded the value of power produced—particularly after energy prices in utility power sales contracts reverted to market prices in the 1990s. Since most of Zond's projects went in the ground later than Kenetech's, Zond also enjoyed the higher subsidized energy prices for the first 10 years of utility power purchase contracts throughout most of the 1990s.

The financing of Sky River, at that time the largest wind facility in the world—was accomplished by Ken Karas, a USC graduate with a theology degree from Ambassador College, an organ of the World Wide Church of God (which

Garner Ted Armstrong made famous through the radio show, "The World Tomorrow"). Whereas Dehlsen was the visionary who got Zond off the ground, Karas engineered its financial success, eventually raising $500 million in institutional project financing. He was, in a sense, there in Tehachapi in the very beginning.

"I remember coming up to the Tehachapi site from Los Angeles in my business suit and tie. It was a Saturday and it was incredibly cold. The wind was just howling. There were just two Storm Master turbines spinning. I remember there was a fine mist everywhere and suddenly a rainbow appeared. Someone took a picture. There I am with a fellow banker and a rainbow and the screaming wind turbines in the background," said Karas. "I thought wind power was promising, but I never imagined that it would grow into a $4 billion annual industry as it is today."

As an officer of the Scandinavian Bank Ltd.'s international finance department, Karas obtained the financing for Zond's first wind farm. He joined Zond in 1982 as its Chief Financial Officer and moved right up the corporate ladder: President in 1986, CEO in 1990, Chairman of the Board in 1997.

Karas made sure Zond did not follow in the footsteps of Kenetech. But there were dark days, especially after the Cashman production tax credits expired. "All of the capital sources that we had relied upon dried up and oil prices that had been projected to be at $45 per barrel and heading north at that point in time were instead $8 per barrel and heading south," Karas reminisced.

"Karas put Zond on a stable financial footing," said Jan Hamrin, who has monitored Zond since that first year she spent in the Tehachapi Mountains. "He took a fresh approach to looking at the costs of wind power. One of the big problems in the industry was that companies were cheating on performance promises. Karas walked into the middle of the debate and asked, 'Why are you cheating yourselves?' The end result of companies promising too much was that the honest wind farmers couldn't get financed. The money guys would say: 'How come it costs you this much when the guy next to you said it cost him only this much?' Karas figured out what was important and how to manage a solid business."

Karas was, and still is, a numbers man. His innovations are less palpable than those of designers and engineers because he deals with the more subtle issues than capturing energy flows with hardware you can touch. Karas was the one who oversaw the development of Zond's strategic partnership with DOE's National Renewable Energy Laboratory (NREL). In essence, NREL helped push Zond's concepts of a new turbine into commercial production by co-funding to pilot projects with utilities in Texas and Vermont. It was also Karas who pushed the company to offer a 750 kW wind turbine—even if it existed only on paper—just in time to win the Minnesota bid and push Kenetech over the edge.

◆

Kenetech's legacy, and many of its machines, have lingered. But those who have worked with Kenetech's five-cent machines (the 33M-VSs) say keeping them running can be hell. Tom Foreman, manager of customer and energy services for the Austin, Texas, based Lower Colorado River Authority, confirmed that the 35 MW wind farm relying on Osborn's nickel turbine performed miserably in 1996. One reason was that the Kenetech bankruptcy hampered repairs and the finding of replacement parts. The next year a storm battered the wind farm region with winds exceeding 160 mph; four turbine towers buckled and 22 blades cracked. Bill Barnes, venture manager for LG&E Power Inc., the largest owner and operator of the 33M-VS turbines, described keeping 200 of these machines on-line as "life in a high-maintenance environment." No answers had yet been found for the cracking of blades, he noted.

Each wind resource region offered different challenges to the Kenetech machine. In West Texas, sudden vertical winds that sweep up a 115 foot rise in the land stress the machines. Lightning strikes are also a hassle. In Palm Springs the multi-directional winds can cause tower cracks. Winds that can jump from 35 to 65 mph in a few seconds and reach as high as 80 mph, inflicting blade problems and "yaw outs." Temperatures that can reach 125°F can damage the internal power conversion components. Rain is a cause of concern, too. "Because of the dry climate, seals can crack when it rains," Barnes continued. In Minnesota, temperatures that can plunge to 20, 30, or even 70 below 0°F, "can keep turbines off-line for several hours in the morning to as long as two weeks." The variability of the weather causes all kinds of problems as visibility can be impaired by blinding snow, which, in turn, can drift and close maintenance roads. Barnes ended on a positive note. "If the 33M-VS can be operated successfully, it bodes well for the wind industry as the next generation of machines hold up better."

Unlike Kenetech, whose turbine represented a dramatic shift in design direction, the architecture of Zond's new machine was quite similar to the Danish wind turbines that the company was so familiar with. Zond was not reinventing the wheel. Although calculated risks were taken, its approach was far more incremental. Nevertheless, Zond could not now afford to have its new machine falter in light of the debacle at Kenetech. "We wanted our Z-Class wind turbines to exceed the kind of rigorous testing and verification that traditional power technologies are subjected to," Dehlsen stated. "We wanted to design the turbine right in the first place so that when we rolled it out the door, our machine was ready for full-scale commercial operation."

Zond set up a test stand in Tehachapi, California, in order to perform "life-cycle" testing of its Z-40 turbine's drivetrain, the interface between the rotor shaft and gears, where the tug of the wind places enormous pressure on a wind turbine. A 900 hp variable-speed diesel locomotive engine that relied on hydraulic actuators was used to simulate the wear and tear a wind turbine endures under the harshest operating environments. Imagine the worst possible hurricane, with wind speeds exceeding 146 mph. That's what this setup was designed to inflict on the new turbine. A shaft from the locomotive's DC motor is connected to the gearbox of the wind turbine and revved up from 30 to 1,200 rpm. Conditions equal to the fatigue a wind turbine endures over the course of 30 years are replicated in just six months. This test stand's prime job was to ensure that Zond's new turbine could respond to stresses associated with rotational drive, rotor weight and yaw, emergency stops, and extreme wind regime conditions. I never saw the test stand in action, but standing next to it I imagined it jostling the huge drivetrain of a wind turbine and I decided it probably wasn't a good place to be hanging around. Even if it didn't exactly simulate Mother Nature, having the stand was a great PR move.

Today, Danish and most other European wind turbines are based on traditional "rules of thumb" design principles that have never been tested. It is common for wind manufacturers to rely on two years of operating history before declaring a turbine ready for the market. This is where the Danes and Zond parted ways. Zond became the first wind manufacturer and developer to challenge these untested "rules of thumb" and systematically examine the real-world stresses on wind turbines. In order to develop the optimal turbine, Zond had to figure out a way to test the stress loads on each individual component of the turbine. This is no small task. Luckily, Zond had already developed the Actual Data Acquisition System (ADAS), which is capable of collecting data at the rate of 160 times per second. The ADAS allows Zond to "properly measure the turbulence that occurs very quickly over a very short period—milliseconds—on each component under a variety of different conditions," said Amir Mikhail, Zond's vice president of engineering. "The key is to isolate places to measure loads," Mikhail added. Zond collected data from over 60 channels, each channel representing an individual turbine component. All told, the data accumulated totaled 1,000 megabytes of information. For items such as the shaft or the rotating hub, Zond relied upon strain gauges and radio transmitters to send data to ADAS.

As turbines grow larger to become more cost effective, and blade spans increase to capture more energy, turbulence stress is amplified. Bigger machines require more, not less, testing of wind load stresses. "There is greater sheer and bending, which results in extreme pressures on the main turbine shaft," noted Mikhail. This is why so many of the early DOE machines failed. In order to address blade stress, the Department of Energy established a blade test facility at the National Renewable Energy Laboratory (NREL) in Rocky Flats, Colorado. Zond's Z class turbine blades not only survived the equivalent of 90 years of wear and tear but broke NREL's blade test stand! The results of this and other testing enabled Zond to redistribute material on the blade in order to better minimize blade stress loads.

Even with all of this testing and analysis, Zond's turbines still revealed glitches when exposed to the elements—nothing quite as dramatic or as frequent as the crumpled blades of the Kenetech machines, but rumors at once swirled that Zond may have come a little too close to emulating Kenetech in cutting its costs to the bone. Zond's design goals now reflected a preoccupation with competing head-to-head with fossil fuels, the mantra of Kenetech before it went under. The following back-and-forth between Randy Tinkerman and Woody Stoddard about the state of the domestic industry in the 1990s applies by proxy to the perceived problems at Zond:

> *Tinkerman:* That we continue to experience catastrophic failures, gearbox failures, blade and rotor failures, ridiculously underestimated O & M, can only point to the fraudulence of the marketplace in which we operate. The margins are so slim (if there are margins at all) that the turbines suffer, even though we know exactly how to build them to higher standards than the automotive industry, even though we know how to build machines that will equal or better projections. The top Euro companies, whose expertise is now unassailable, and the sole U.S. player are forced to bring turbines to market prematurely simply to keep up with Internet financing delusions and an energy market that historians will look at as criminal, as civilization destroys its own children.
>
> *Stoddard:* We now have the instrumentation that can actually track the tip path of individual rotor blades, in real time. What can this do? Nothing short of preventing a tower strike in real time. Furthermore, this sort of failure prediction can be done with fatigue monitors and onboard signature analysis. Can we afford this?
>
> *Tinkerman:* Of course we can't afford this, which is precisely the problem. Renewable energy is far more valuable, *economically*, than what the market currently pays. I'm tired of watching companies who evolved from the days when we didn't know how many months before the next design crises would appear, who survived more than a decade (and some two) of life on the financial edge, who somehow maintain the current level of engineering sophistication, yet still have mid-'80s type operations problems. Not because they don't know how to build the machine, but because they are forced to play the energy cost shill game.

The utility owners of the first two Z-40 installations were far more forgiving than Tinkerman in their assessment of contemporary wind technology. The first installation was in 1995 in the Fort Davis Mountains of west Texas, for Central & Southwest (CSW). The small 6.6 MW wind farm relied on the first-stage prototype, the blades of which featured ailerons similar to the flaps on an airplane wing. These ailerons are intended to maximize power production by changing pitch in response to changes in wind direction. Only the CSW-owned machines feature these ailerons; they were abandoned shortly thereafter. (The Z-40s installed in Vermont pitch the entire blades.) Ward Marshall, a program manager with CSW, was confident that despite some start-up problems, the Zond machines would ultimately work well. "The Danish are much more conservative in their design changes," Marshall noted. He said he would give any of the Danish machines the edge right now in a head-to-head comparison with U.S. wind turbines.

Green Mountain Power (GMP) owns the second project using the Z-40 turbine, which was installed on the Searsburg Ridge in 1996. GMP has been quite happy with the project. Though some bugs had to be worked out, including the replacement of 2,600 bolts inside the rotors in the dead of winter, the machines have performed up to expectations given their prototype status. "Some things are less than optimal because there's not a lot of field experience [with the Z-40]. Zond hasn't had a lot of time to debug and refine the design," said Bill Ralph, GMP's senior planning engineer. During the first four months of operation, however, the turbines were able to produce power over 97 percent of the time, under harsh winter conditions. One of the main goals of the GMP project was to test performance during cold winters, when the winds blow strongest in Vermont and demand from customers for electricity is high. Ralph worked with Enron Wind Corporation (EWC, as Zond was renamed in 1997 after purchase by a Houston firm) to customize the turbines for Vermont's winters. For example, the three rotor blades are covered with a slippery black surface to minimize ice accumulation and to concentrate the sun's energy to shed ice. Heaters and synthetic Teflon-based lubricants also enable the turbines to continue spinning at temperatures as low as minus 40°F. Finally, tubular towers were used so that field workers can service the turbine by climbing stairs inside the tower, safe from the sting of the elements.

The "paper" turbines are now real turbines up and running in Minnesota. Dehlsen and Karas won a second contract to build yet another 100 MW-plus wind farm with its Z-750 model, proof that Karas's bold strategy to capture America's largest wind projects had worked. "We didn't think we could be competitive with the Z-40," recalled Dehlsen. "It was Ken's idea to bid the Z-750. He had the courage of conviction to bid the project with a machine yet to be." Dehlsen had just returned from the grand opening of the Northern States Power project and he was choked up. "I had this vision such a long time ago about how wind power was such a logical thing. I thought it would be ubiquitous. But after suffering so many tough blows, you begin to doubt yourself. What I saw in Minnesota was the embracing of wind power in the Midwest—the place which has the best wind resource. I also saw the budding of a new constituency for wind power farmers—which can help put political pressure in Washington, DC to really take advantage of wind power."

Chapter 24

Moving Mountains with Wind Chicks, Churches, and Oil Companies

It is your own conviction which compels you; that is, choice compels choice
—Epictetus

The wind farms that went into the ground in the early 1980s began a process of dismantling the electric utility monopolies that had dominated energy supply decisions since shortly after the turn of the century. As the wind farmers and other independent energy producers plugged their power into the utility grids created by Edison using Tesla's AC current, they began to take control of our energy future away from the largest electric utility monopolies in the world. While Kenetech worked in collaboration with utilities such as PG&E, most wind companies prided themselves in their independence, often kept at the expense of a hardscrabble existence.

The low-price bids for new wind and other independent power sources developed under the BRPU in the early '90s convinced large energy consumers, such as cement kilns, that California, whose rates were 50 percent higher than the rest of the country, needed to introduce more competition into power supply decisions. Beginning in 1994, the California Public Utilities Commission began a long, tortured debate over how best to restructure the state's $23 billion power market. It came up with a plan in December of 1995; then the California legislature took over and unanimously passed AB 1890 at the end of the legislative session in 1996. The end result: Most California consumers have been able to choose power suppliers, much as we pick long-distance phone companies, since April Fool's Day 1998.

The virtues of a free market, and allowing customers to choose their electricity suppliers, would be tested in California, the state that has been on the forefront of designing energy policies for quite some time. Since polls showed 60 to 70 percent of consumers across the country preferred renewable energy—and about 20 to 25 percent were willing to pay extra for it—California would be a major test to see whether these impressive numbers could be translated into market outcomes.

Free market enthusiasts predicted there would be a mad rush to purchase the cheapest power possible, and a new kind of company, the power marketer, began to entice the state's citizens to switch electricity sources. Some predicted half of California customers would switch. But they were way, way off. Only 1 percent of the eligible customer base switched to any alternative supplier during the first year. Still, these customers represented 11 percent of the state's total consumption, reflecting the fact that the large electricity consumers dominated the early exodus from incumbent utility providers. A promising sign for wind and other renewable resources was that virtually all of the residential customers in California who did switch switched to a green power product.

Over the course of the first year most of the marketing to residential customers in California shifted from trying to sell cheaper to trying to sell greener. Nationwide the early green marketing campaigns generated responses from roughly 1 percent of potential customers. The high visibility of a wind turbine on a bluff outside of Traverse City, Michigan, boosted this community's participation rate to more than 3 percent, one of the highest success rates of any green marketing program in the country. This admittedly small success nonetheless illustrates wind power's primary role today in generating support for the infant green power market.

As wind farms surpassed 10 years of operation, many were in a state of decline and decay. Projects that had received as much as 11 and 12 cents per kWh at the end of 10 years of operation, based on Eckland's oil forecasts back in the late 1970s, suddenly were receiving only 2 to 3 cents per kWh. The low energy pricing that had hit Eckland first was now crippling the rest of the California wind industry. Because power surpluses suppressed prices, the market rate for electricity was also quite low in the late '90s. Some of the AB 1890 funds are dedicated to keeping these existing renewable energy facilities alive, but many of them are expected to shut down if this transition fund runs out in 2002.

Instead of a guaranteed market, wind power now has had to compete with other conventional sources of electricity and with other renewables. There was no longer one buyer—the utility monopoly—but a wide range of potential customers, from residential consumers, who knew nothing about electricity, to large industrial firms with their own energy management staffs. In a world of customer choice, the big question was whether Californians and consumers in other states would voluntarily send their electricity dollars to machines that worked less than a third of the time, allegedly killed eagles, and were first erected with the help of the last great American tax shelter.

Interestingly, wind power emerged as the poster child of green power marketing in California. "Wind power has captured the imagination of the American public," confirmed Jan Hamrin, now executive director of the Center for Resource Solutions (CRS), whose office is located in San Francisco. Having quit the Independent Energy Producers years ago, Hamrin has developed a successful energy-consulting career and now travels around the world promoting renewable energy technologies. "People can visualize wind turbines. It is the one renewable energy technology they can see in their mind. Wind turbines have emerged as the preferred power source of those wanting to buy green. And they are now used as a selling tool, with a special emphasis on constructing new wind power facilities." This former wind farming crew camp cook argues, "There's a market for wind power in a deregulated world. A wind company can respond rapidly when there is a need for power. This is particularly the case when one is building out an existing wind farm and one goes through the regulatory hoops in advance."

"Deregulation is offering customers a choice, and much of the general public wants wind," confirmed Julie Blunden, marketing manager for GreenMountain.com, a Vermont-based company that has set its sights on California's small consumers. In its first year of competition, GreenMountain.com offered three successive green power products. The highest priced, "Wind For the Future," captured about a quarter of its residential customer signups. Considering that its price was 19 percent above the cost of generic power, GreenMountain.com was quite pleased with the response. "It's like when you are at the store and you have three options and you just decide to go for the best," said Blunden. "Wind power has a wonderful personality. It is so visually interesting and people really get it. It is a customer-friendly technology that you can add in small increments. It is more expensive that way, but it can be done. You can't do that with a geothermal power plant, which has to be built in 50 MW chunks at a time. Besides, it is hard to visualize a geothermal plant in a simple picture. And solar still seems too far away. Wind power just happens to be in the sweet spot."

Only 10 percent of GreenMountain.com's "Wind For the Future" product was from new wind facilities. For every 3,800 customers who signed up, the company put up a turbine. Two were erected in 1998. On the first anniversary of the opening of the California market, the company offered an upgrade, Wind For the Future 2.0, which boosted the "new" wind portion of the product to 25 percent. It also lowered the price, reflecting the growing competition among green power marketers for new customers.

AB 1890, California's landmark restructuring law passed in 1996, introduced a half billion dollars in new subsidies to help the renewable energy industry cope with drastic changes in the way electricity was being bought and sold. The first new California wind farms in over a decade were being built, in the Palm Springs area. These and other renewable energy projects total 483 MW. One of the main lessons learned in California was that a variety of obstacles still keep individual consumers from seeing the environmental impacts of their power purchases. For example, cumbersome procedures for switching suppliers can delay deliveries of green power for months, frustrating consumers. More formidable was a temporary surcharge on all consumers' bills, which could be more than a third of the electricity bill, to pay for past utility debts, including investments in old nuclear reactors and power purchases from wind turbines. This surcharge reduces any real electricity savings until after 2001.

GreenMountain.com has had better luck in signing up customers in Pennsylvania, where the rules of the game starting on January 1, 1999, made it easier for customers to switch. The new market structure means people can actually save money by switching. Instead of California's 1 percent of consumers, some Pennsylvania utilities have lost a quarter of their customers to the competition. GreenMountain.com grabbed 29 percent of those who switched in the early going, a percentage matched only by the incumbent utility's lowest-cost offer. In June 1999, GreenMountain.com announced plans to build what it billed as the largest wind farm in the eastern United States, a 10 MW facility located in southwestern Pennsylvania. The wind farm came on-line at the start of the new millennium and will serve green power customers in the Keystone State and New Jersey, which also offers consumers a choice in power supply. Over 40 states in the United States are in the process of opening their electricity markets to competition. What California and Pennsylvania tell us is that wind power is emerging as the technology of choice for those concerned about the environment.

◆

One way to address some of the skepticism of Californians about buying wind and other renewable resources is to work with community nonprofit groups to build knowledge and support for clean power through old-fashioned grassroots organizing. That's the approach used by Rudd Mayer, green power analyst for the Boulder, Colorado, based Land and Water Fund and self-described "Wind Chick." An heir to the Oscar Mayer family fortune, an avid skier and fan of the Grateful Dead, Mayer is the type of person an infant green power market needs. She combines the down-home genuine dedication of a committed tree hugger with the instincts necessary to cut deals with utilities and all kinds of electricity consumers. Her raspy voice—a result of damaged vocal chords—only helps her charm her way into the green power market's inner sanctum. She is highly sought after because she helped launch the nation's most successful wind power marketing program in Colorado. As of December 1998, Mayer had helped aggregate 17 MW of subscriptions for new wind power, with subscriptions coming in at roughly 1 MW per month. Of this total, 9 MW came from residential customers, 5 MW from businesses, nonprofits, and governmental entities, and 4 MW from wholesale customers, such as rural cooperatives. No other green marketing program has achieved as broad a participation profile from such a diverse customer base. Over 350 businesses, at least eight local governments, numerous hotels, churches, and colleges, have all signed up. Grade schools hold wind power penny drives. Mayer has even convinced builders and developers to offer new homebuyers a year's worth of wind power.

"I'm not selling so much a product, I'm selling an ethic. In 100 years of turning on the lights, this is the first time people here can say, 'Make my power from the wind, don't burn any coal,'" is the way she sums up the Land and Water Fund's fundamental message. Mayer is one of those unique individuals who feel equally comfortable at a rock concert handing out flyers and in the posh corporate boardroom of Coors Brewing Co., one among a number of high-profile corporations she has sold on wind power.

Unlike California, Colorado has not yet restructured its electricity market. Mayer is selling wind power for the Public Service Company of Colorado (PSCo), a utility monopoly, under the name "Windsource." PSCo did not have high expectations for Windsource, whose supply comes from new wind turbines installed in northeastern Colorado. Thanks to community-based grassroots marketing, sign-ups doubled initial expectations—from 5 MW to 10 MW—just within the first six months. One reason for the high participation rate is that purchasing wind power in Colorado is far less complicated than switching to green power in California. Windsource sells wind power in blocks of 100 kWh for $2.50. The switch is immediate and, in most cases, requires no complicated calculations. For an average residential customer, the extra $2.50 per month (the equivalent of a 5 percent premium) would supply 20 percent of the home's electricity from the wind; for $12 extra per month, a typical home could be completely powered by clean electricity from new wind farms.

Mayer explains the discrepancy between California and Colorado: "Customer acquisition costs are too high for suppliers to reach and educate customers by themselves. An environmental group can serve as a catalyst for a series of public policy and private actions in the community. The nonprofit partner can lend credibility to the product and marketing message. Partnerships between suppliers, in this case PSCo, and the nonprofit community, as well as state and local governments and businesses, can reduce customer acquisition costs by five times when compared to traditional marketing techniques utilized by private power suppliers."

While the Land and Water Fund (the "LAW Fund") focused on a grassroots strategy, it also took a much broader-based approach than those initially employed by power marketers in California. A public relations coup occurred early in the program when then-Governor Roy Romer of Colorado agreed to purchase enough wind power to supply the entire needs of the Governor's Mansion in Denver. "Windpower was part of my childhood experience. I'm from the eastern plains of Colorado, and we used to use wind generators on our farms in the 1930s because we were not hooked up to the power lines," said Romer. "I believe each of us can have a role in expanding the use of these future-oriented technologies and take advantage of Colorado's plentiful supply of wind and sun. Renewable energy such as wind power is a promising resource whose time has come in Colorado," said Romer.

Perhaps one of the most instructive efforts was to engage corporations and local governments as Windsource "champions," with each buying 250 100 kWh blocks (roughly 15 percent of the output of one wind turbine) for a contract period of three years at a cost of $25,000 annually. The cities of Boulder and Denver became champions, with Boulder purchasing enough wind power to supply half of its main municipal building. Boulder linked its Windsource purchase to new investments in energy efficiency, whose savings will, according to city staff, completely offset the cost of the clean power investment. Most of the initial champions were large corporations: Coors, U.S. West, CF&I Steel, and IBM.

The largest consumer of wind power in the United States. justified its investment by the dramatic reductions in CO_2 realized by switching from coal to clean power. The New Belgium Brewing Co. of Fort Collins, Colorado, announced in late February 1999 that it would go well beyond any other Windsource champion by purchasing 1.8 million kWh of wind power per year, about the same amount of electricity as produced by a single 660 kW wind turbine erected near Medicine Bow, Wyoming, in fall 1999. The brewer made a 10-year commitment to pay an annual premium of over a half million dollars per year, which will add 20 to 30 cents to the cost of each barrel of beer produced at New Belgium.

"Beer breweries produce a tremendous amount of CO_2 as a natural byproduct of the fermentation process required to make beer," said Greg Owsley, marketing director for the company. "The industry is now looking at ways to capture and reuse the CO_2. Since we purchase most of our electricity from coal plants, we realized we could replace all of the CO_2 with a wind power purchase. It was an easy sell," said Owsley. He noted that all 70 employees voted unanimously to purchase wind power even though the purchase would shrink company bonuses, which are based on the brewery's production costs per barrel of beer.

New Belgium staff discovered that the coal burned to generate electricity for the beer plant emitted four times more CO_2 into the atmosphere than brewery operations themselves. With the wind power purchase, New Belgium has decreased CO_2 production by over six times the amount that the CO_2 recovery technology would have. Each single wind turbine will reduce the amount of coal burned by over 980 tons per year, eliminating over 4 million pounds of CO_2 per year.

New Belgium has incorporated the wind power purchase into its marketing plans. The brewer periodically produced paper coasters that can be serve as post-cards. It released one such coaster postcard that featured beer bottles for wind turbine towers and announced that New Belgium is the first wind-powered brewery in the United States.

The individual responsible for siting and permitting both Pennsylvania's GreenMountain.com and Colorado's PSCo wind farms is no stranger. His name is Dale Osborn, and he is current president of his own company, Distributed Generation Systems, Inc., based in Evergreen, Colorado. He's a one-man show now, and he seems to relish the role of soothsayer despite his Kenetech baggage. Osborn continues to see parallels between the energy and computer businesses, a legacy, no doubt, of his days with Texas Instruments. "The visionaries that spurred on the computer revolution are poised to enter the electricity business in a big way." He claimed confidently that they would replicate the success of distributed PC computer networks by fostering a market for distributed wind, solar, and other micro power generation systems: "Every shopping center, every corner store will have the ability to self-generate in the next year at a cost of 5 cents per kWh. Utilities are going to wake up one day and find out that 20 percent of their best customers are gone. Why? Because they can't supply the quality of power industries such as the semiconductor requires. Before 2005, large segments of the power market will be off the system from a peak power standpoint. They will be relying upon distributed generation sources." Osborn spat out this scenario as if reading the script for a documentary movie.

Osborn was hired by PSCo to help permit the new wind farm because he knew about the bird issue: "We screened ten to twelve sites and about half passed our tests. If we found even one eagles' nest within a mile of a proposed site, we excluded it from consideration." He sees the future of the U.S. wind industry as being in small projects—10 to 25 MW—with very short construction time lines based on short-term contracts pegged to premium priced green energy. Both the PSCo and GreenMountain.com projects fit that succinct summary. "I'm just trying to push the technology forward in any way I can," said Osborn, acknowledging he is also pushing some small fossil-fuel-powered units. But he did a bit of bragging, claiming that the total development cost of the PSCo "green pricing" wind project was just $110,000. A representative of the Colorado utility remarked that Osborn knew how to put together deals and that his claim that his new market model was "brutally inexpensive" was no lie.

With no overhead, little or no up-front fees, and an ability to keep firm deadlines in order to get paid promptly, Osborn has found his niche in the brave new world. Interestingly, he now uses Danish wind turbines for his projects.

◆

One of the most interesting strategies for signing up consumers for wind power in California involved Episcopal churches, in an arrangement with GreenMountain.com that could become a model for the rest of the country and is already being replicated in Iowa, Massachusetts, Pennsylvania, and a growing list of other states. Propelled by daunting issues such as global climate change, and responding to a new trend of environmental activism within the Bay Area's religious leadership, St. Aidain's Episcopal Church, in San Francisco, became the first church in California to go green. It signed up for the "Wind for the Future" product in July 1998. Sally Bingham, chair of the Commission for the Environment of the Episcopal Diocese of California, has led the crusade for churches here to switch to green power. She prepared a resolution, entitled Episcopal Power and Light (EP&L), which was adopted in October 1998 by the statewide diocese. It instructs each Episcopalian church in the San Francisco Bay Area to buy clean, renewable power. Why should churches care about issues such as renewable energy? "God's purpose for us is to love and to live in harmony with all that God has made. All of Creation and all generations to come are our neighbors. As we take God's word to heart, we are assuming a leadership role in the healing of our planet by putting our church on an energy diet. We invite people of all faiths to join with us in cutting greenhouse gas emissions by investing in energy efficiency and by buying renewable energy resources generated from God's gifts, the wind and sun."

"It is not a historical role for churches to get involved with issues like green power. But churches are filled with people with enormous passion and that believe there is something out there in the spiritual realm," said Tom Bowman, of St. John's, in the Oakland hills, one of the several Bay Area Episcopal parishes that is buying a portion of its electricity from new wind turbines "If we are to ever witness a sea change when it comes to the environment, churches have to become involved." Unlike corporations, government, or academia, other institutions that can sway public attitudes, "churches reach ordinary people. As I often say to fellow parishioners, 'If not us, who?'"

A key goal of Bay Area Episcopal churches is to sign up 3,800 parishioners. When that goal is reached, Bingham hopes to erect a new wind turbine that will have exclusively been supported by members of the Episcopal faith, a first among the nation's religious community.

◆

The problems, both real and perceived, in California's infant green power market are not surprising. Just as the birth of modern wind technology was the result of an odd marriage between idealistic engineering visionaries and greedy money merchants, the success of a green power market will ride on the shoulders of confused consumers and sometimes clueless private power marketers. Many of the consumers most concerned about global warming and sustainable development are also the folks most leery of private enterprise, a paradox the green market will have to overcome if it is to fulfill its great promise.

Most environmentalists, with the exception of the Natural Resources Defense Council and the Union of Concerned Scientists, have not fully embraced the idea of green power. Since we never see electricity—only know that it is not there during a power outage—calling this virtually invisible substance "green" raises a number of important questions. Most important is how we know the electricity is green if we never get to see it.

The "Green-e" Renewable Electricity Branding Program was launched through CRS in the fall of 1997 to address that very issue. The brainchild of Jan Hamrin, the Green-e program has been touted as the nation's first voluntary certification and verification program for "green" electricity products. The centerpiece of the program, the Green-e logo, is displayed on promotional materials from power marketers offering products that meet ethical, consumer, and environmental protection criteria. The Green-e logo program was modeled on the success of the recycling logo in establishing the credibility of "green" consumer products. Companies using the Green-e logo to denote a specific product will verify its renewable energy content, abide by a code of conduct governing their business practices, and provide customers with regular information about the sources of the electricity they purchase.

Despite these tracking and certification systems, the concept of green power has been dinged quite heavily. The Washington, DC, group Public Citizen issued a scathing analysis of California's green power market in late 1997. Among the complaints lodged in its report, authored by Nancy Rader, former consultant to AWEA and Zond's ally in 1989, is that there are significant opportunities to circumvent and abuse the aforementioned certification and verification processes. Though the concept of "green" power in the abstract is quite appealing, transferring the concept into the reality of the complex electricity business has not been easy. As with any new business, it takes a while to get all the bugs out.

Given the impact of critics such as Rader and confusion and misinformation about wind and other renewable energy sources, environmental groups have not launched an organized effort to push the purchase of green electricity, and so by default private corporations are emerging as major players in building demand for new wind farms. A case in point is the Ventura-based Patagonia, Inc., an outdoor clothing company well known for its commitment to the environment. In July 1998 Patagonia became the first commercial customer in California to purchase 100 percent of its electricity from a new renewable energy project, a "merchant" wind farm near Palm Springs being constructed by Enron. This and plants like it are called "merchant" facilities because, unlike plants of the past, their proprietors are not relying on a utility power purchase contract but instead are betting that there will be enough consumer demand to sustain them. Interestingly enough, the power consumed at Patagonia's 14 facilities roughly equals 1 million kWh per year, which can be easily supplied by one of Enron's 750 kW wind turbines.

The process of choosing a clean power product was far from easy, according to Jill Zilligen, Patagonia's director of environmental programs. "I thought it would take 6 months. I was naive. Instead, it took 14 months," she said. One critical problem was that none of the four power marketers that responded to Patagonia's Request for Proposals had yet developed a commercial product; initial green marketing efforts had been focused almost exclusively on residential customers. "There was an awful lot of back and forth," she acknowledged. Patagonia set a very high standard. The company preferred a 100 percent new renewable source located in California as close as possible to the firm's Ventura County headquarters. The new Enron wind farm, a three-hour drive away, was the only offer that met all of Patagonia's preferred green product criteria.

Patagonia "wanted to show how committed we were to the environment" by choosing the greenest of the green products. What about eagles and other birds, did they factor into your decision making? "We were not thrilled about the unfortunate side effect of contributing to raptor kills—and we did weigh that in our decision making—but the consensus was that greenhouse gas emissions from nonrenewable electricity sources were a far greater problem." Zilligen did quite a bit of research on the topic and was not convinced that avian mortality was a major problem.

Making commitments to wind power in light of what has been perceived as a negative attribute of the technology, bird kills, illustrates how corporations can play a major role in legitimizing wind and other renewable energy sources for small consumers. "Corporations are important to building the green power market, from a volume and educational standpoint. We can also send the message to residential customers that it is not difficult to switch," said Zilligen. She does not see the wind power purchase providing any near-term sales advantage or other immediate competitive edge to Patagonia. "We instead looked at the long term, when resources become scarce. We will then have a competitive edge because we will have reduced our use of nonrenewable sources and will be as efficient as we can. Part of our justification for paying the premium for wind power was that it would provide a new incentive to be more efficient. We have already done quite a bit of energy efficiency upgrades, the easy stuff like better lighting. So we are now dedicated to becoming even more efficient. We may not entirely offset the wind power premium, but hope to come close."

◆

Patagonia purchased its wind power from turbines manufactured by Enron. The burden of underwriting the development of its new line of wind turbines had come under tremendous scrutiny because of the blemished track record of Kenetech. In the mid-'90s Zond was forced to seek long-term solutions to its growing financial needs. The climate still wasn't right for an IPO—Kenetech alone probably sealed the fate of that option—so Zond began thinking about potential joint venture partners. Ironically enough, it found what Ken Karas had said was the perfect partner: an oil and natural gas company whose new strategy was to become a wholesale supplier of all forms of energy, including electricity, all across the country. The purchase of Zond by Enron in the spring of 1997 was a move that raised many eyebrows both in the United States and elsewhere. For some, the purchase was proof that wind power had indeed come of age, for even companies that made a great deal of money from oil and natural gas were now seeing a bright future for renewable energy. For others, the Enron purchase was a cynical attempt by a company spending $200 million in national advertising to cloak itself in the virtues of green energy while continuing to peddle more-polluting forms of electricity to most of its customers.

The jury is still out. But Karas claims not much changed after the Enron purchase. "And the changes that have happened have been for the better. Enron has huge capital reserves and is well-positioned in key power generation markets around the world. Sure, they were a gas company. But they want to be a diversified energy company now. Their primary goal is to be the largest seller of electricity in the world. We fit pretty well into their product diversification and differentiation strategies," Karas said. Among the key advantages of hooking up with Enron is that a firm with such a broad presence in the market could help "firm up" wind power when it is offered to customers. Having a power marketer as a strategic partner could accelerate the integration of wind into power supply portfolios. With the help of Enron's marketing skills and access to low-cost capital, Karas hopes to transform the wind operation into a billion dollar per year business by 2002.

Even Woody Stoddard, the wind industry's long-time curmudgeon, is impressed with Enron. "They hired me because they were having '80s type problems. I was viewed as the wise old sage. Kenetech poisoned the well and I was hesitant to get back involved with wind. But now, it's getting very interesting. There is a whole new generation out there."

"Kenetech kind of bullied its way to the top," Stoddard claimed. "What I saw in Tehachapi is that they [Enron] are allowing good people to do good work." Enron also purchased Tacke, a German wind turbine manufacturer. Stoddard claims he could see the influence of Tacke on Enron's approach to design and engineering. "Tacke started out as a family-owned operation with clunky machines. But the Germans have the manufacturing experience to build big turbines. They are capable of building turbines that only require maintenance twice a year—and will last twenty." For comparison, he noted that Kenetech's workhorse machine—the 56-100s that were derived from his initial design—

required maintenance every 60 days. "Young people in their 30s—very creative people—are being given the resources that enable them to make a career out of wind power engineering. In Germany, it is the cream of the crop that want to develop wind power. A parallel situation in the U.S. would be the space program, which is where the nation's elite engineers all ended up." He continued, "What Zond has is the field experience—that's the edge they have on the Germans. You can't overestimate just how important that field experience is. I think they are coming up with a new machine that will be able to compete head-to-head with big European machines. What I saw was innovation after innovation. Unlike Kenetech, Zond has learned its lesson."

Chapter 25

Landscapes of Power

Places privileged by nature have been cursed by history.
—Eduardo Galeano

"I cannot comprehend a process which allows people hundreds of miles away to use the resources of other people in a destructive manner. . . . Do people really need electric shoe buffers, toothbrushes, garage door openers, can openers, and hundreds of other modern conveniences? Do people really prefer living in pollution?"

These are the words of Peterson Zah, a long-time leader of the Navajo Nation, having served as president and chairman throughout the 1980s and 1990s. Zah made these comments a long time ago. He was caught between members of the Navajo tribe who wished to integrate into American society through "good" jobs in the coal industry and those who feared the ultimate doom the mines might bring to the reservation. Maintaining balance here, at the spot where Arizona, New Mexico, Utah, and Colorado all come together, has become a real-world energy challenge.

Taken as a group, 45 percent of Native Americans live below what the federal government defines as poverty. Unemployment here in the Southwest's scalding desert hovers around 35 percent, but is less severe than on many other reservations—where it can reach 90 percent. And jobs associated with coal are a large part of the reason. Many Navajo once thought the discovery of coal, as well as uranium, on their land was a miracle, since this, the largest Indian reservation in the country, was once described as "the most worthless land that ever laid outdoors." Some 70 percent of the work force in the coal mines on Navajo Nation land and at the Four Corners and Mojave power plants is Navajo; the mines and plants provide 50 percent of Navajo tribal revenues. These workers report hearing loss, blindness, and shortness of breath. Mysterious deaths, and even sheep born without heads, have prompted many tribe members to re-examine their dependence on coal for their livelihood. In the past 30 years more than a billion dollars of coal has been gouged out of the earth at Big Mountain, which sits on top of one of the world's richest coal deposits, in northeastern Arizona. Each year, Big Mountain yields 12 million tons, 5 million of which are pumped through an underground slurry line to the Mojave Generating Station in Laughlin, Nevada—275 miles away—where the coal is burned to generate power for most of the Southwest and Los Angeles.

"We can no longer see the beauty of the land the Great Spirit put here," Zah says. "Now is the perfect opportunity to shift gears and take a new direction. We have the space, the people, and the land. What we are now doing is going to be our downfall." The Navajo and Hopi Indian reservations underscore the poverty of this country's energy and natural resource policies. Millions of Americans rely upon electricity fueled by natural resources found on their lands. Yet large numbers of Native Americans, including the Navajo, have no electricity for their own needs and suffer the health consequences of business-as-usual energy production and pollution.

Americans derive over half of their electricity from coal. The popularity of this solid fossil fuel is largely due to its low price. Yet this price reflects major loopholes in federal air quality laws and ignores some very real and unacceptable environmental costs. Many coal-fired power plants, for example, were exempted from key provisions of the federal Clean Air Act, originally passed in response to the first Earth Day, in 1970. It was assumed these dinosaur power plants would be shut down after 30 years of operation. Today, because of more active trading of electricity across state borders in an era of competition, many of these power plants are running more often—not less.

Even though virtually no coal is burned in California to generate electricity, the power that most California consumers have been buying over the past few years keeps getting dirtier and dirtier—and coal is the culprit. How is this possible? Unlike other generic products we buy in the store with labels detailing ingredients, our day-to-day electricity fuel mix can have wide swings due to seasonal availability of hydroelectricity and availability of nuclear power plants. No set content criteria apply to the generic power that any consumer gets if he or she does not switch to a new power supplier. Whatever is sold into a central state power pool that day ends up powering your home or business. Coal-fired plants from other states contribute to that pool.

Since California deregulated its market in 1998, there has been progress on the green power front, but there has also been a distressing increase in consumption of coal, which jumped from 17 to 20 percent of the state's overall fuel mix in 1999. Most of this electricity is imported from western coal plants, like the Mojave generating station in Arizona. The federal Environmental Protection Agency (EPA) claims Mojave is the single largest source of pollution, robbing future generations of the historic, pristine vistas of the Grand Canyon.

Nationwide, coal power plants are responsible for 93 percent of the sulfur dioxide and 80 percent of the nitrogen oxide emissions generated by the entire electric utility industry. These emissions spawn acid rain, which kills forests and fish, and create urban smog, which has been linked to respiratory ailments. Seventy-three percent of the carbon dioxide emitted into the atmosphere from electricity generators comes from coal-fired power plants, increasing the threat of global warming.

The list of ill environmental effects doesn't stop there. The mining, processing, and transporting of coal also have impacts. In the West, about 87 percent of coal is removed from the earth through strip mining, which can contaminate soils with heavy metals and destroy near-surface aquifers. Coal combustion also results in huge quantities of solid wastes, often tainted with toxic residues. Enormous quantities of waste heat generated by coal plants require large amounts of water for cooling. Collecting this water from lakes, streams, and reservoirs often threatens local aquatic life, killing fish as they smash into screens designed to keep all organisms out of the power plant.

◆

"Every time you flip a switch, you are helping eradicate Navajo people," charged Marsha Monestersky, consultant for the Sovereign Dineh (Navajo) Nation and co-chair of the Human Rights Caucus for the United Nations Commission on Sustainable Development. "The United States likes to point its finger to human rights violations in other countries, but never to itself."

Here is what just a few Navajo revealed in a secret report prepared for a California utility about what it is like working at, and living near, the coal mines:

> Before it rains you can see the coal dust on the plants. The animals eat this and get congested. They have runny noses, they cough, they have film over their eyes. . . . We are concerned because we see black spots on our animals' liver when we butcher them. Young lambs have yellow and black spots on the liver and inside the intestines. Red spots and lines, too. Our children get bronchitis a lot. Even the babies. . . .

Another retired Navajo coal worker said this:

> I started working for the mine in 1973. I have Black Lung and silicosis. I have no appetite and can only eat certain foods, mainly vegetables. I shouldn't even drive a vehicle because I black out. My roof needs to be fixed and I need running water and electric-

> ity so I can have an indoor bathroom, refrigeration and have electricity to run breathing equipment that I need. We are living near the mine, but have no electricity.

Perhaps the most egregious of insults is the moving of the graves of the Navajo elders. According to Navajo spiritual beliefs (which are still followed by the Dineh, a shrinking pool of Navajo traditionalists), there is no prayer for those who have been reburied. Many have never even been informed where the remains of their loved ones have been moved. As they sacrifice their burial sites and ranching land to coal mining, the Dineh and other Navajo clans fear that they are witnessing the expiration of their traditional culture.

One former federal official who was partly responsible for the current hard feelings between Indians and Anglos has led the charge to replace existing sources of economic development with new, cleaner approaches that complement the spiritual beliefs and promote sustainable management of natural resources. His name is Stewart Udall, former head of the Center for Resource Management (CRM), a nonprofit organization affiliated with movie actor and environmentalist Robert Redford. Udall, brother of former Arizona Senator Morris Udall and Secretary of the Interior in the Kennedy administration, traces his family ties to early Mormon settlers in the Southwest. In 1993 Udall headed a conference entitled "Why Renewables? Why Now?" in Santa Fe that focused on an opportunity presented not only to local Native American tribes but to the rest of the West as well. "Interest in renewable energy at state and regional levels is being revitalized by new and expanding markets, increased diversity and flexibility, and concerns over managing the environmental and political risks of conventional energy sources," Udall said in an invitation to the conference. Energy leaders in the West "are exploring new mechanisms for cooperation" to develop these resources, he noted. The meeting, attended by 60 high-level decision makers, was the first step in what has become a regional effort to develop consensus on how to phase out coal and nuclear plants and bring eco-friendly options, such as wind power, on-line. Udall has a fondness for wind. "Inventors and technicians will have to pay homage to windmills. They'll have to build machines that use, not abuse, the unearned gifts of nature," he once said.

The task won't be easy, but Udall carries some important weight behind him and has a humanitarian reason to push alternatives to business-as-usual energy policies. When he was Secretary of the Interior, Udall approved the Peabody Western Coal Company's strip mining on Big Mountain, a spot considered sacred by both the Hopi and Navajo tribes. This is one of the largest coal strip mines in the United States. This and other mines were "sold" to the Indians, and the general public, as necessary economic development. The generous terms for the coal companies in their contracts deserve scrutiny. One contract with the

Navajos, for example, paid out a royalty of 15 cents per ton of coal, a sum that is one-tenth the $1.50 price applied to non-Indian land!

The Hopi and Navajo Nation lands, rich in uranium, coal, and groundwater, are also home to the fastest growing population of Indians in the country. The Navajos, which number 200,000, surround the Hopi, which number 4,000. What was described as an ongoing land dispute between the two tribes has played into the hands of mining interests. Some Hopi have gone on record to say that the mining operations should be shut down despite the fact they derive income from coal mined on their lands. In contrast, many Navajos still view the mines as one of the few sources of jobs in the area.

The Hopis practice a special two-day ceremony every summer that speaks directly to the stewardship issues raised by the prospects of wind power and other solar energy sources in holy landscapes such as theirs, now so heavily scarred by the coal companies. This annual ritual, as I described it in a report for the Center for Resource Management, aptly summarizes choices that face native tribes—and societies around the world:

> Brightly colored Kachinas, which represent everything that is perfect, appear in the town plaza. Some represent clean water; others, clean air. The clowns, who are actually men covered with mud, find this wonderful Garden of Eden unexpectedly. The Kachinas are solemn, but the Clowns are oblivious as they sing nonsense songs. Suddenly, the Clowns discover the Kachinas. They rush to them and try to stake out their turfs. The Clowns shove and push and fight. They eventually calm down and begin entertaining with satirical skits about ordinary life, about the ineptitude of government, about lewd and vulgar acts.
>
> Meanwhile, the Kachinas come and go. Shortly after lunch, however, an Owl appears and silently circles the plaza. Few heads turn as it completes its survey. Nevertheless, the next time he appears, the Leader of the Clowns is waiting. The two meet inconspicuously. After a brief chat, the Owl leaves again. Upon his third visit, the Owl meets with the Clown Leader for a second time. The Kachinas keep dancing; the Clowns keep clowning. The Leader reaches into his pocket and offers the Owl corn meal, then a turquoise necklace. But the Owl, with Large Unblinking Eyes, shakes his head, No. Instead, the Owl lays out whips of different lengths. Some are made from Yuccas, others from different types of trees.
>
> According to this Hopi theater, the Owl is asking the Clown to choose the appropriate punishment. He must accept responsibility for not getting his House in order. (According to Hopi beliefs,

> Owls are messengers for the Gods and issue warnings about ecology.) He must accept some pain for not inspiring his people to heed the call to sustain the sacred landscape. Of course, the offerings by the Clown are attempts to buy out of the penance, an approach the Owl rejects.

Currently, about 3,000 mostly elderly Navajos who live in abject poverty are being forced to move from Big Mountain. About 13,000 Navajos have already been relocated. Congress passed the Navajo-Hopi Settlement Act in 1974 to resolve what had been described as "range war" between the two tribes. Under the act, land surrounding the Hopi reservation that had been used by both tribes became strictly Hopi territory, and Navajos were forced to move out. But critics say the alleged range war was largely a ploy designed to open the land for coal mining by affluent Hopis. The Hopis were supported by the Peabody Western Coal Company, whose long-term goal of expanding strip mines was served by the approved partition scheme. "There is evidence that Peabody hired a public relations firm to promote the range-war story," claimed Gabor Rona, an attorney with the Center for Constitutional Rights, a New York-based public interest legal organization that has argued that relocating Navajos under the 1974 Settlement Act is illegal.

Against the backdrop of the "energy crisis" of the early 1970s, and convinced that the two tribes were at war, Congress swiftly passed the relocation act. The Navajos and Hopis have never been buddies (one Hopi word for Navajo can be translated as "head bangers"). Yet, for the most part, they peacefully coexisted for hundreds of years, until fossil fuel was discovered on their land in the 1950s. The tribes never really fought until tribal governments were established to negotiate mining leases with coal companies.

More than half of the power generated by the Mojave plant goes to Southern California, but Southern California Edison, which was half owner and operator of the plant, has never claimed any responsibility for the impacts on local Native American communities. Edison's share of the plant was purchased by AES Corp. in May 2000, but there is little reason to expect that this independent power developer will deviate from business as usual in failing to address the social impacts of coal operations on local tribes.

Luckily, Harris Arthur, a Navajo energy consultant, sees a prime opportunity to shift gears toward a more sustainable energy system today. "Some 25,000 Navajo, and another 25,000 other Native Americans, do not currently have electricity," he noted. Remote Indian villages are often miles from transmission lines. Photovoltaic systems, which convert sunlight directly into electricity, and wind turbines can be installed directly on homes without the need for imported and polluting fuels or power lines crisscrossing the landscape. These solutions to energy woes honor both past cultural practices and future needs. While some Hopi tribes refuse to have power lines on their land because they believe the infrastructure scars the land, they are enthusiastic about small, off-grid solar systems. Indeed, the few that currently enjoy solar electricity offer prayer feathers to their solar panels just as they do to crops, rivers, and the land.

Tribes such as the Navajo and Hopi view the harnessing of solar and wind power as one of the few business opportunities that is reconcilable with their legacy of self-determination, sovereignty, and environmental values. Both traditional Navajo and Hopi tribal cultures invest immense spiritual power in land and regard the earth as a living being. There is a new movement, applauded by a growing list of environmentalists, for these and other Native Americans to return to their spiritual roots. Because Native American lands such as those owned by the Navajo and Hopi feature the best solar and wind energy sites in the country, among the most economical power sources for these Indians will be renewable energy systems.

◆

Stoney Anketell is a wind farmer for the new millennium. He believes the wind, considered a sacred force among his Native American ancestors, can indeed bring prosperity. While born on Fort Peck tribal land in Montana, Stoney was raised in the southwestern part of Washington State, near the mouth of the Columbia River. He was a tennis professional and was earning decent pay as a tennis teacher when his back went out. His brother, who had a seat on the Fort Peck tribal council, then convinced him to come back home, where he too joined the council. Stoney quickly became immersed in the oil business, of all things. "While I was on the council in 1996, we approved the largest oil exploration deal—in terms of total acreage—of anywhere in the United States," he said. The deal was with Gulf Canada. Shortly thereafter, Stoney became intrigued by wind power. "It is clean and attractive. It didn't take much to convince me that wind power is the future. Unlike fossil fuels, you don't have to kill off the planet in order to turn on the lights," he said, his eyes twinkling through his glasses.

Stoney believes wind power represents the most attractive opportunity for members of Native American tribes to become major players in the new, deregulated power market.

According to a report published in early 1997 by Udall's Center for Resource Management, tribes have the legal ability to build power plants and transmission lines and to deliver electricity at the retail level. No other creature of governance has as many options for energy management as do the tribes. "If people are willing to pay more for green power, they might be even more interested in purchasing green power from Indian tribes," comments Paul Parker, author of the white paper. "If tribes were focused on the issue of renewables, they could become charismatic leaders for the entire nation, bringing their moral and historical weight behind a national effort to choose clean power."

Anketell wants to put Parker's ideas into practice: "I want to sell Native

American green power." He has been working to install a wind farm in northeastern Montana for quite some time. He is among those who view wind power as an answer to the threat of continued pollution, and as a new source of revenue for tribes. Bechtel Corporation has just completed a year-long wind resource assessment of prime reservation wind sites, reporting that several, on land owned by the Assiniboine and Sioux tribes, are the cream of the crop for abundance and strength of the wind. Anketell claims the results of the $250,000 assessment justify a 45 MW near-term goal, a 200 MW mid-range goal, and ultimately a 500 MW wind farm—the same capacity as a major California wind farm region. Transmission access for an initial phase of development is readily available. The problem has been finding a customer.

Anketell maintains it is only a matter of time before wind machines start showing up on the high desert plateaus and mountain canyons of the Rocky Mountains, the Great Plains, and the Southwest. There are no real alternatives: Coal plants are unattractive because of air pollution; new hydroelectric dams are out of the question because there are already so many and they kill fish; and new nuclear plants are prohibitively expensive and produce yet more nuclear waste for which there is no federal repository. "What else can they do?" Anketell asks. He is currently working to establish a nonprofit organization to help educate Montanans about renewable energy. "We are hoping to get our first wind turbine up soon and then hook it up to a hydrogen fuel cell to store the power," he said enthusiastically. Since residents of Montana are not allowed to choose their own power sources until 2002, he knows his efforts will need to be sustained for quite some time. "We are going to have to develop the green market ourselves because the local utility monopolies are not interested. And we are so remote out here. But the general public wants green power and we hope to deliver, first to Montana, but ultimately to other places—Native American green power," he concluded.

◆

The work Anketell did with the Fort Peck tribes grew out of a grant from the Department of Energy, which offered a pot of money to Indian tribes to investigate energy resource development under the Energy Policy Act of 1992. Though most tribes used the money to conduct feasibility studies, one recipient, Marty Wilde, used it to erect what he claims is the first wind turbine on Native American lands.

The installation of a 100 kW Danish wind turbine in May 1996 was largely a public relations ploy to get the Blackfeet tribe excited about the prospect of large-scale wind power development. "We placed the turbine on a hill at the Blackfeet Community College near Browning, the Blackfeet tribe's headquarters. It sits right dab in the middle of the best wind site in the lower 48," said Wilde. It is projected that 10,000 MW of potential wind power could be developed here—enough electricity to serve the needs of more than a handful of states.

Wilde notes that while some locations have higher average wind speeds (annual readings here average 20 to 22 mph), no other location boasts so large a potential wind development arena. The installation, he says, is "a glowing example of how local people took the initiative. Historically, hustlers have promised the world to these tribes, only to let them down time and time again. This project could be a major moral boost that will allow the Blackfeet tribes to determine their own destiny." While the wind resource on land owned by the Blackfeet tribe is outstanding, getting the power to customers poses a challenge. The Blackfeet tribe lands border Glacier National Park, a pristine Rocky Mountain destination that, though more remote than Yellowstone Park, ranks as one of America's top tourist regions. The huge park stands in the way of what could be cost-effective transmission corridors to deliver power to urban consumers.

According to Wyatt Rogers, a consultant to the Denver-based Council of Energy Resource Tribes, one of the bright spots for wind developers is that "the fastest growing U.S. power markets are nearby." He noted that Seattle and the rest of the Pacific Northwest make up one of the most attractive, and growing, electricity markets. Rogers, a Native American himself, is optimistic about wind power precisely because such renewable resources "fit in with our traditional philosophy. Sources of natural energy that can be regenerated are definitely preferred over sources that must be wasted."

Robert Gough, secretary of the Intertribal Council on Utility Policy, is even more gung ho, particularly about wind power development in South Dakota, a state whose wind resource is fantastic, but which receives little attention. "Tribal lands in all of the Great Plains could help develop a red variety of green power," said Gough. "Most people don't realize that Native Americans are ten times more likely to not have electricity than the average American," he continued. Largely this is simply because many Native Americans live in isolation. On the plus side, from an electric supply point of view, is that Native Americans tend to be distributed in denser communities than other consumers living throughout the Great Plains, Rocky Mountains, and Southwest. "We are very concerned about global warming. Many hunters tell me they can already see the changes occurring when they are out in nature. The cycles are different than they used to be. We own millions of acres and are strategically located to also deliver power to the Chicago and Ohio River valley, regions which are considering closing a number of nuclear power plants," said Gough, whose long, graying ponytail reaches all the way down his back.

All told, the federal government recognizes the sovereignty of over 500 different American Indian tribes and Native Alaskan groups. Nearly all have established land holdings, independent tribal governments, and a growing demand for more energy to fuel emerging and rapidly expanding economies. Today, tribal memberships are growing at an average annual rate of over 3 percent, which makes tribes the fastest growing demographic group within the United States (outside of immigrant populations). Tribes living on remote lands represent the largest group of people not currently enjoying the convenience of electricity or telephones.

The history of energy resource development on Native American lands is fraught with tales of exploitation. Native Americans were typically resettled on reservations whose land was viewed as largely uninhabitable, lacking water and having poor soils or harsh elements—including wind and sun. These land holdings trace their roots to the days of Kit Carson, General Custer, and the soldiers and cowboys who "won" the West.

Renewable energy development could represent a break from the past and, unlike coal, uranium, and oil reserves, will not steer all of the profits to large corporations. "Over the long term, renewables are symbolic of sovereignty and self-determination," Parker said. While outside expertise is necessary, Native Americans themselves need to play a large role. "In the past, tribes have been passive. They need to be aggressive in order to take advantage of the limited window of opportunity that exists with deregulation."

Chapter 26

A White Knight for Dark Times?

It's not windy—we are just blowing you a kiss.
—sign at the entrance to San Gorgonio

I am missing the opportunity to drink some fine Taittinger Brut 1995 Napa Valley sparkling wine and instead am climbing a 200-foot-tall wind turbine tower on a dare by Lu Setnicka, the PR boss for Patagonia. A former Boy Scout and football player, I could not run. But like a fool, I forgot to wear gloves, and I'm gripping the ladder intensely as I go higher and higher. Thank God the wind isn't too bad this day in the San Gorgonio Pass. In fact, the wind is coming out of the east instead of the west, the normal direction for the Santa Anas that flow in from the Pacific Ocean, almost 200 miles away. These wind conditions occur on only a handful of days every year. It is December, normally among the quietest for energy production in most wind farming regions in California, including this one.

I get 40 feet up and imagine being Terry Mehrkam opening up the guts of a wind turbine at this height and then falling to the ground. I get up to 80 feet and think about my buddy Bill Graham—no, not the dead rock promoter—and how he lost his job at Kenetech a few years ago and now gets most of his dough from inherited stocks in oil companies. I keep going up, pausing for the first time when my legs start to shake a little. I'm wearing a hard hat and, like a rock climber, am connected to a safety harness. Just the same, my signature was required on a liability waiver.

I've been attending the dedication of "Green Power I," a new wind farm that Enron Wind Corp. claims is the world's first "merchant" wind power facility;

that is, the first large wind farm built to serve private-sector customers rather than to fulfill a power purchase contract with a utility monopoly, like the ones U.S. Windpower, Zond, and Jan Hamrin relied on in the early '80s. A transformation is underway, but not within the realms of finance and public policy. What is most evident to me here, in San Gorgonio at the dawn of a new millennium, is the sheer grandeur of today's wind power technology. Enron's new machines, and the Danish NEG Micon recently installed by SeaWest, look like the future. In this sparse land of long, dry horizons, the turbines at times have the appearance of military sentries. It must be the straight and even rows on the flat floors of the desert champaign. Cabazon, the wind industry's most disconcerting site, captures the changes. Smaller, older machines, most of which never worked, are being replaced by these newer, bigger, and better turbines that really do. "The area has been a wreck since 1985," acknowledged Bob Gates, one of the original members of Zond, now a senior vice president at Enron Wind and recently elected to a one-year term as president of the American Wind Energy Association. "Since it is the first thing you see on the highway when you come into the San Gorgonio wind resource area, cleaning Cabazon up gave me great satisfaction. Instead of having 230 wind turbines stuffed into the site, there are now only 22—and they are sleek, state-of-the-art machines."

I had planned to go on what had been billed as a "Champagne Bus Tour" of Cabazon. The event was, more or less, a toast to the wind industry's seeming and sudden maturity. Cabazon, where flying turbine blades chased a fleeing teenager worker and a falling tower nearly killed a crane operator, was once like wind farming with the Marx Brothers. Today, the imposing white machines of Enron, arranged in a staggered array on an uneven platform, looked like swans flying in tight, geometric formation. Or perhaps a dance troupe, or ballerinas, engaged in a mesmerizing and perpetual pirouette. But Lu from Patagonia, whom I just met in person for the first time, keeps egging me on. "If I go on the tower, you have to go to. It's a once in a lifetime experience," she insisted. I later discovered Lu was a rock climber and was the first female to climb these towers.

She went up before me, right behind Hollywood actor Ed Begley Jr., who first invested in wind turbines nearby in 1985. A photographer with a telephoto lens is snapping shots of Ed's head, peeking out of the nacelle at the very top. He and Lu were both so far up that I could barely see them. Though the tower is 40 feet wide at the base, it narrows to eight feet at the top. One has the odd sensation of falling backwards the higher one gets. At this height—the equivalent of a 20-story building—the sense of scale hits home. This single piece of machinery, with its 76-foot blade locked and stationary me, produces enough electricity for 750 or so homes. When I finally climb all the way to the top, I am pooped.

Apparently there are still some good stories from the field going around. A windsmith sitting above, at the hole in the bottom of the giant nacelle, told me about a fellow windsmith at another site who was inside the rotor cone when the turbine was energized and the blades started to rotate. "He kept throwing out his tools to try and signal someone below to stop the damn machine," laughed the ponytailed narrator, his shades hiding his eyes, his mouth forming a wide smile. "Finally, somebody saw all of his tools and shut the machine off." Another windsmith, invisible to me because he is sitting on top of the gigantic rotor shaft that intersects the nacelle, chimes in, "He wasn't hurt too bad. Just a few bumps and bruises." They both laugh loudly.

Things have been pretty peaceful in these parts, was the report I gathered from these and other windsmiths and a few other Enron personnel. It wasn't always so. The only other old-timer at the Green Power I event, Zond photographer Lloyd Herzinger, pointed to a spot over toward some ridges to the northwest. "That's where the dude from Southern California Edison was Victor Morrowed," he said, a not-so-oblique reference to the decapitation of a Hollywood actor I only remember seeing in the old "Combat" TV show in the '60s. Herzinger's comment reminded me of a story that I had thought was just an ugly rumor. One of the early SCE vertical-axis prototypes, quite similar to the one that blew up at the first wind power conference in Palm Springs in 1981, self-destructed during a dedication—and actually beheaded one of the ceremony's presenters.

As I sat up in the nacelle, I was a little too tired to fully enjoy the view. Still, there was ample evidence everywhere that the wind industry had come a long, long way. I was standing there on the ladder with Steve Kelly, executive director of the Renewable Energy Marketing Board (REMB), an arm of the Independent Energy Producers, the group that Hamrin started in 1982. REMB had won a contract to spend ratepayer dollars doled out by the California Energy Commission to educate consumers about the virtues of green power. The fact that money was being spent to entice consumers to bolt from their incumbent utilities and buy from firms offering wind and other renewable resources seemed significant.

When Steve and I finally made it down to the ground, the drunks on the bus were returning from Cabazon. At the curb were a bunch of unopened, chilled bottles of champagne. Steve and I grabbed a couple before heading back to the airport. I knew one thing. They certainly didn't have ceremonies like this back in the '80s.

Earlier that day, before I had been talked into climbing this tower, an Episcopalian priest read a blessing from Psalm 104 of the Hebrew Scriptures. "Ride on the wings of the wind," is the only line I remembered. "Wind was a messenger," remarked the priest, Father Andrew Green. "We live in a glorious world that is in concert with the divine, but which has been tarnished by fossil fuels," he said. "We need to take a stand on a clean environment. It is God's creation."

Then there were the 175 giggling, shy, and rowdy kids from Two Bunch

Palms Elementary School, who were barely able to squeeze in under the large circus tent that was put up to protect everyone from the wind and sand. Each child had participated in a wind power poster contest. Patagonia had printed up shirts with the winning design for all of the kids. "One hundred percent organic cotton," Lu pointed out. After singing the national anthem, they launched into an original rap tune entitled "Windmill #5." I couldn't tell you many of the words except for the chorus, a sing-song repeat of the single word "spin."

Mary Bono, who took over her late husband Sonny's congressional seat for the Palm Springs area in the mid-'90s, boasted that she was the original sponsor of the latest reincarnation of the federal wind energy production tax credit, which had just gathered enough votes to extend a 1.5 cent per kilowatt hour subsidy for another five years. The strangest part of the program, however, was Sean Lomax, a world grand champion whistler with a three-octave range. A short, stout man, he wowed the audience with his true and occasionally adventurous renditions of old standards such as "This Land is Your Land."

◆

While the ceremony for Enron Wind's new wind farm implied that wind power technology was booming, there were also signs that some of the old issues that haunted the industry would not go away. At about the same time that Enron was touting its new wind farm for green power customers, the Audubon Society held a news conference in Washington, DC, to denounce a proposal by EWC to build a 40 MW farm at Gorman. This was the same spot in the Tejon Pass, between Bakersfield and Los Angeles, where in 1989 the Audubon Society joined the Sierra Club and the Tejon Ranch Co. to kill a proposed 77 MW wind farm, almost sinking Zond in the process. Now, 10 years later, the same battle is being fought. The earlier proposal had been rejected because of fears of what might happen when California condors that had been raised in captivity were released into the wild. Now the wild population is up to 49 birds, with 20 of them living in the vicinity of the proposed wind farm.

"It's hard to imagine a worse idea than putting a condor Cuisinart next door to a critical condor habitat," Audubon Society vice president Daniel Beard announced. "We believe this project must be stopped. Enron is proposing to build a death trap." Also at the press conference was a representative from the Tejon Ranch Co., whose 270,000 acre contiguous land holding is the largest private real estate parcel in California, and which the proposed wind farm site would border. Tejon linked up with the Audubon Society and helped bankroll its campaign against Enron, this time thrusting the issue into the national limelight. The company paid for ads in Washington, DC, newspapers denouncing the proposed wind farm and for billboards in Los Angeles and Enron's home city of Houston featuring an image of a raptor heading into wind turbines, with the caption "Kill the Condors?" Also underwritten by Tejon is an Audubon website devoted exclusively to killing this wind farm.

Tejon's cattle operations provide about 60 percent of its revenues, and the company touts its properties to film studios as ideal locations for movies set everywhere from deserts to forests. But the company sees its future in real estate development, and Tejon has long worried about what wind turbines might do to plans to build posh homes and condos in the area. Tejon's concern for condors also seems to be highly selective: In 1997, the company filed suit to halt condor reintroduction in the Tehachapi Mountains, fearing it could hamper its plans for development.

The company's concern for condos, however, is unwavering—and may point to its real reason for fighting turbines. In an interview, Tejon general counsel Dennis Mullins made it clear he is no fan of wind energy. He described the potential wind farm near Gorman as "a visual blight 12 o'clock high at the gateway to Kern Country." But sprawling development is apparently no visual blight in the eyes of Tejon. The company had recently struck a deal with three top U.S. homebuilders to develop 4,000 acres adjacent to the land near Gorman that Enron had been eyeing, and plans for a huge mall were also in the works.

Then a very odd thing happened. A couple of months after that press conference, Audubon's Dan Beard had changed his line. "Wind power is a clean, nonpolluting source of electricity, producing no acid rain, oil spills, or radioactive waste," he said in early November 1999 when he announced that EWC would develop the 40 MW wind farm at another site on land owned by the Tejon Ranch Co., some 22 miles away from the original site, in a forlorn part of the Antelope Valley. "Enron Wind has a long history of dealing with environmental concerns in a positive, responsible manner. . . . [I]t is a company committed to protecting the environment and the region's wildlife."

"The Gorman experience was useful for both of us," Beard continued. He noted that Audubon and Enron Wind pledged to jointly screen any sites for bird populations of special concern before future development could proceed. While recent studies by the CEC and DOE imply that the threat to birds from wind turbines has been greatly exaggerated, Beard insisted that the Audubon Society "has a responsibility to go slow, to be very careful and thorough in our analysis of potential wind farming sites." Freely acknowledging that the aggressive DOE goals to meet 5 percent of U.S. electricity demand with wind power by the year 2020 "give me a little heartburn," he added that he is an optimist and believes that "we can protect migratory birds and develop alternative energy sources such as wind power." He concluded, "Audubon does not want to be Chicken Little crying 'the sky is falling, the sky is falling.' We also need to work towards solutions."

The Enron people were probably foolish for attempting to build a wind farm at Gorman. The negative press attached to its 40 MW proposal for Gorman sends the wrong message to the general public. "No one seems to pay much attention to the number of birds that are fried by transmission and distribution lines that transport power from nuclear and coal plants," commented V. John White, executive director of the Center for Energy Efficiency and Renewable Technologies. A key difference between wind power companies and other providers of electricity is the former's swift response to concerns about birds. "The wind industry is sinking considerable resources into seeking solutions to the avian mortality issue, while the nuclear industry, after decades of operation, has yet to come up with a credible plan to protect citizens from the very real dangers associated with radioactive waste," White said. "When one steps back and takes a good look at the big picture, the level of bird kills coming from wind turbines is completely dwarfed by the looming catastrophe of global warming and air pollution impacts associated with the present status quo energy picture."

◆

Ken Karas finally called me back, once the latest chapter in the Gorman story ended on a positive note. He had more than a few surprises for me.

The first surprise was that Enron Wind had a new wind turbine, designed with a little help from Woody Stoddard. It was based largely on a design by Tacke, the German manufacturer that Enron purchased. These machines were twice the size of the last model Enron delivered to the market. At 1.5 MW, they were larger than the Smith-Putnam turbine and roughly the same size as the DOE monsters. Enron had an even larger machine—2 MW—that was specifically designed to be installed offshore, much as Heronemus had originally envisioned. One of the only aspects of the original Zond turbine incorporated into the new TZ 1.5 MW were airfoils developed with DOE money; these maximized energy capture by changing the profile of blades while a Zond controller/inverter helped maintain utility-grade power from the ever-variable wind.

"I thought 500 kW was about as big as wind turbines were going to get," Karas said, reflecting on the recent dramatic increases in wind turbine dimensions. "I know we will never see a 100 MW wind turbine, and I feel pretty comfortable that we'll never see a 10 MW turbine either. Over the next 10 years, I think the size will probably settle around 3.5 MW. Anything bigger than that, and you simply can't move the stuff around on highways, let alone get the turbines to the installation site." (In fact, Enron is now building a wind turbine that could be as large as 5 MW—some four times larger than the Smith-Putnam machine.)

Still, Karas would not rule out unforeseen developments with wind power technology. "Did you know they wanted to close the patents office in the early 1800s? They thought everything that could be invented had been invented already. Obviously, that was not the case. As materials science continues to evolve, and all of the innovations occurring right now in power electronics and the new ways of manufacturing gear boxes, it is hard to say how it will all settle out," said Karas.

Karas then turned philosophical. "This is a very different kind of industry. It is very fun stuff. I'm looking outside right now here in Tehachapi. The wind is only blowing at 12 to 13 mph and the sun is going down. The wind turbines are moving slowly and it feels very soothing. It is really hard to feel bad about working in the wind industry."

I found out two days later that Karas left Enron Wind. The company decided it wanted to own 100 percent of the company, so they bought out Karas's minority share. Like Dehlsen (whose remaining share Enron also purchased), he was now quite rich. Yet another one of the major figures of the early wind farming days was moving on, a sign of the times, of the maturing of the wind farming industry as it joined the corporate mainstream.

Chapter 27

The Promise

> The substance of the wind is too thin for the human eyes . . . written language too difficult for human mind . . . spoken language mostly too faint for human ears.
>
> —John Muir

The last time I drove through the Altamont Pass, I had an urge to knock on the doors of the few folks that actually lived out here in this wind. I wanted to hear their stories. Daryl Mueller, one local, did speak to me. He is part of the anti-wind power crowd that successfully stalled projects in Washington and Montana in the mid-'90s. Mueller moved to the Altamont in 1986, onto a piece of property that lies right across from Kenetech's very first wind turbine installation. Though he did not witness the blade throws, he can't stand wind farmers. Hates their guts. A noisy nuisance and big taxpayer boondoggle, he'll tell you. "They want to export this rotten, 'environmental product' to third world countries, a technology that will just cut their raptors to shreds," complains Mueller, who is a general contractor by profession. He thinks new coal plants are the answer. He also has a soft spot for nuclear power. "You know Three Mile Island, most of it was just a hoax, a mud slinging contest conducted by the media. It's a shame how this country has discredited nuclear power."

Local rancher Joe Jess likes the wind farmers but is pissed these days because bankruptcies and technical glitches with the Fayette turbines on his property have him making a lot less money than some of his neighbors with well-operating machines. The income from wind farms has kept the ranches here in the Altamont alive. As companies go under, and as prices paid for this clean electricity plummet as they track low fossil fuel prices, ranchers here start cussing like hell. But in Minnesota and Texas, farmers are emerging as the strongest supporters of wind power because the revenues from farming the wind are saving family farms.

On this drive through the Altamont, I come upon an ancient water pumper, a relic juxtaposed against a hazy backdrop of smog on this hot summer day, brought to life by spinning white propellers. I then head east and after rounding a sharp curve suddenly see one golden eagle—an immature bird—heading north. Then two more drop in unison from the lattice tower of a huge transmission line. As I head uphill, all three immature eagles soar over my truck one after another.

Then I come around another curve and the view is incredible. It is about 5:30 in the afternoon. Long shadows give the thousands of white spinning pinwheels depth. They seem to be chanting: A low resonance is audible, a sound frequency not unlike the chants of Gyuto Tantric monks. I pull over and sit for about 10 minutes and then continue on, following the curves of the road until I come upon the Tesla substation, a grim reminder of the ugliness of large-scale infrastructure required for transporting massive amounts of electricity. I have never before seen such a display of intense mechanical and almost otherworldly industrial equipment in such a bare setting. There has been no attempt to hide Tesla. Just a short distance from the row after row of gray, metal structures lies the Altamont Speedway. Cars whiz by, their occupants anxious to catch tonight's race. My guess is that they don't think much about the turbines. They have become part of the landscape here.

◆

With the 1997 purchase of Zond by Enron, a growing global giant whose prime business has been producing and selling natural gas, an era ended. Jim Dehlsen was relieved of his duties as chairman of the board. He received the tidy sum of $40 million, and he and his wife packed their bags and moved from the redneck sticks of Tehachapi—from their ranch constructed of recycled wood across the street from the cement kiln—to sunny Santa Barbara. It was there, near the Pacific Ocean and his sailboat, that Dehlsen had first begun visualizing the wind reaping machines. Approaching 60 years of age, and the ultimate wind industry survivor, Dehlsen was trying to slow down. He set up his own consulting business with Brent, his son with a fresh MBA. The primary client of Dehlsen & Associates was initially Enron's wind operation, which he started up.

Jim Dehlsen went to Kyoto, Japan, in December 1997 to represent AWEA in the negotiations for a global warming treaty. Some 5,000 representatives from 160 nations were meeting there to negotiate probably the single most difficult agreement in human history. In a way, Dehlsen's life had come full circle. The burden of keeping Zond afloat in the incredibly rugged energy economy had been lifted, and he now had a new lease on life. He described what it was like in

Kyoto. "The fossil lobby plastered the meeting areas with these slick presentations questioning the need for any change in government policy. They vastly outnumbered any other single industry there. Representatives of the nuclear power industry popped up all over the place too, painting themselves as the only logical 'green' alternative to business-as-usual energy supply." Surprisingly, Dehlsen acknowledged some sympathy for the captains of the fossil fuel industry. "They had a job to do—to bring energy to the nation. They had the right values, they built up solid businesses. And now, for the first time, they had to justify their existence."

First and foremost, Dehlsen is an entrepreneur. He even respects the fossil fuel competition, despite their large environmental footprints, because of they've managed to grasp success, so elusive in the market. However, he told me he thought the roles between the fossil and wind power industries were flipped at Kyoto. "The perception among many in the electricity industry [in the years before Kyoto] as well as the general public, is that the wind industry was nothing but a fraud. Wind turbines only worked because of the tax credits." Now, instead—at least it seemed to Dehlsen—it was the fossil fuel industry that was being painted as the impostor. The lavish subsidies afforded the fossil fuel industry were finally being acknowledged, probed, and questioned. "We are currently consuming 175 million barrels of oil per day. Fifty metric tons of oil. If you wanted to visualize the planet's exhaust pipe for CO_2, what would it look like? The surface area would be nine square miles and the column of gas would rise 3,300 feet into the atmosphere. Can you imagine the dark film surrounding the planet after 40 years of that kind of abuse?"

◆

"The wind industry is no longer about technology," Randy Tinkerman stated without hesitation. "The technology appears to be proven. And we have crossed the hurdle in lowering costs. Today, it's all about politics. The main issues in wind energy have nothing to do with wind energy. The issues revolve around corrupt energy policy in America, the most notorious aspect being energy pricing." Tinkerman observed that fossil and nuclear lobbies have not only distorted energy pricing through lavish subsidies, but are thwarting efforts to develop a coherent national response to the global climate change threat. "It is a travesty," he said. "The insurance industry—which has already quantified the effects of global warming in their actuarial charts—is now focused on how to pass these costs onto the consumer rather than solving the problem. That's a travesty too."

Without strong public policies in support of renewable energy technologies such as wind power, global climate change will continue to heat the planet despite all the forward-looking eco-volunteers like Patagonia and the New Belgium Brewery. Their actions should inspire, not obviate the need for, stronger policies promoting all renewable energy sources. "Our climates are the great sculptures of the natural world," Ty Cashman told me. Working to convince Japan to enter the wind and renewable energy business in a big way, Cashman always takes the long and broad view. "We have already endured five great extinctions. We could be witnessing the sixth—the first perpetuated by humans. Because we live in our climate-controlled houses, a lot of us don't really know how we are fucking with the climate. We need to think about future generations. If there is a God, imagine him judging the legacy we have left for future generations. As one looks out at great vistas from the mountaintop, what would one think. We as a species inherited a paradise but left behind a wasteland. It's not too late to shift, but time is definitely running out."

Since the termination of the generous federal and state tax credits in the mid-'80s, Bergey Windpower, which still tops the small turbine market, has focused on export markets. "In this country, a small 10 kW wind turbine might reduce your electric bill by $100 to $125. It doesn't improve the quality of life a whole lot. But in places like Morocco—where they carry water for miles every day—a 10 kW wind turbine can pump enough water for 2,000 to 3,000 people. Wind power can also cut back on disease because families don't have to rely upon contaminated surface water." Working overseas, nonetheless, has its own ups and downs. Bergey related what happened once in Morocco, when Ramadan, the month-long Muslim holy period of fasting, began right in the middle of an installation. "We were viewed as infidels since we continued working. Our Muslim driver would just scowl at us. He would stop his vehicle, place his prayer rug on the side of the road, and pray for ten minutes. But when the installation was complete, the driver told everyone in the village that they no longer had to travel every day to get water. It was hard to keep a dry eye."

Bob Sherwin, a pioneer whose Enertech Corp. went bankrupt in 1985, continues to market smaller wind turbines through a company called Atlantic Orient, headquartered in Vermont, not too far from the remnants of the Smith-Putnam machine in the Green Mountains. "What kept drawing me back in are the people. They are world-class. Sure, there are some crooks and assholes, but you find those kinds of people in any business. You know I gave up a teaching career and took a big hit—it's hard. But there is nothing like the feeling of bringing power to a small rural village. And seeing the sparkle in the eyes when they get electricity for the first time in places like Kazakhstan, where the old Russian electric grid might operate for 8 or 10 hours a day, or in Alaska, where the wind turbines lower the fuel costs of Native Americans."

Small wind turbines are an excellent alternative to business as usual on Native American lands—as well as in developing nations. Over 150,000 small wind turbines have already been installed around the world, and that number could easily be doubled or tripled within the next few years. At the 10 kW level of power supply, the only power source cheaper than wind is micro-hydro systems. The future market for small wind turbines is huge, since two billion people currently lack electricity. Extending the electric grid to all of these far reaches of the planet would cost more than $1 trillion, an investment that would carry an almost unimaginable environmental price tag as well.

The United States has retained a global leadership role in appropriate-scale technologies that can deliver power without environmental insults. American firms dominate global markets for small wind turbines used as off-grid, stand-alone, or hybrid power supply systems. These modern relatives of the unique American water pumpers of old have captured 60 to 70 percent of the global market for small wind turbines, and the annual growth rate of that market has exceeded 20 percent for the past several years.

Many Americans also enjoy living off the land with the help of wind power. Lawrence Mott, for example, relies on the wind to supply 90 percent of his electricity for his large, beautiful home overlooking a lush meadow surrounded by dense hardwoods in Vermont. Many of the deciduous trees are maples—he lives in the heart of Vermont's maple syrup country. Bergey might feel that small wind turbines don't add much in the way of quality of life here in the United States, particularly from a monetary perspective. But folks like the tall and lanky Mott groove on the psychic thrill of not being plugged into the local utility's grid. "I live with my wind power, scheduling parties to coincide with winter winds," said Mott as we onlookers gazed up at his wind turbine, its tower poking through a canopy of trees. Mott works for Northern Power Systems (NPS), yet another pioneering wind power company that hails from New England. He got the 1 kW turbine for a low price since it had a small defect that created a little extra noise, which didn't seem to bother anybody. Mott pointed out that in this region of Vermont, there were no zoning laws so he didn't have to deal with any government red tape to get his wind turbine in the ground. "This is America at its finest," he proclaimed.

Mott's employer, NPS, was once purchased by John Kuhns (the same person who bought Eckland's Fayette) when New World Power was busy buying up firms left and right before all of the money raised in its public offering was frittered away. One may frown on Kuhns's investment ethics and questionable business tactics but he did tell me a number of things that still resonate as a fundamental truth: "Village power—integrated systems that feature wind and solar photovoltaic systems—is where it is at. Village power systems can take advantage of the distributed nature and the beneficial character of the wind and sun. But village power systems will not happen until utilities get into a deregulated frame of mind."

Carl Weinberg, one of the nation's gurus of what are called "distributed generation" power sources, agrees with Kuhns that the future lies in developing wind and other power sources that are small and modular and can be sited close to the point of actual power consumption. It could take a while before this happens, but it could also happen a lot quicker than a lot of utility executives think.

"It took bath tubs 25 years to reach a 25 percent penetration of the population. For VCRs and cell phones, widely considered some of the fastest moving products ever, it took 8 years to reach the 25 percent of the total market figure," he added. Over the course of the next decade, the emergence of a market for distributed wind turbines in this country and particularly in the developing world could be huge. But if we are indeed serious about tackling huge problems such as global warming, village power won't be enough. Given the huge pile of statistics that has accumulated demonstrating the folly of our continued reliance on fossil and nuclear fuels, it has become clear to me that we need to think small and local, but we also need to think big and global. In other words, we need to simultaneously develop small wind turbine markets as well as continue to pursue the installation of the giant wind turbines that Heronemus once envisioned.

But there also appears to be a future for the big machines. Today's state-of-the-art wind turbines are approaching the size of the Smith-Putnam machine. In 1999, America's largest wind turbines went in the ground, appropriately enough, in Texas. Generating 1.65 MW of electricity, these Danish Vestas V-66 wind turbines are not only larger than the Smith-Putnam but also taller than the Statue of Liberty. The towers of these gigantic machines are over 270 feet tall; the tip of the blades reaches 370 feet in the air! One could deduce that Divone was right.

Divone readily admits that many of the DOE-supported wind turbines did not perform well. "All along, I thought it would take three generations of technological progress to produce commercial wind turbines." To Divone, investments in gigantic machines were the logical choice to make given the information available at the time. The Smith-Putnam turbine had worked, so he and many others at DOE saw no reason to shy away from extremely large wind turbines. In his mind, DOE was playing the role the federal government should play when commercializing a new electricity-generating technology: funding projects that nobody in the private sector could afford and, in the process, generating research results that could help the private sector not repeat the same mistakes made by the federal government.

Yet Divone's math, and all of the sophisticated turbine designs drawn up by what some claim were the nation's second-string aerospace engineers, delivered machines that were never appropriately scaled to the state of knowledge about the stress the wind can impose on blade, tower, or rotor. All of the sophisticated calculations failed to account for a very steep learning curve required to develop a machine that could always adequately respond to the diverse stresses supplied by fickle winds. In a sense, the federal approach to developing wind turbines failed to recognize how the raw elements of nature impose their fierce will upon machines when they move from paper to the harsh, real environment.

The most disappointing story of all in the history of U.S. wind farming, however, is that of the Carter father and son team. These two stubborn Texans have nothing to show today for a fundamental wind turbine design that folks like Woody Stoddard, one of the most respected wind turbine engineers in the wind business, swear is the where the technology will be—not next year, but early in this new millennium. "The Danish machines are a dead end," Stoddard said. Yet he did not minimize the challenges facing such a lightweight machine. "Carter's flexibility puts a very large burden on engineering and control. It's like using a high-performance, light Americas Cup yacht—designed for the Pacific—in the Atlantic, instead of using a slow, heavy one. The sleek one's OK for the Atlantic, but only if you know exactly all the things that could happen out there. But if we can control and make Carter blades reliable, they definitely hold the key to the future."

Stoddard is not alone in his glowing assessment of the contributions the Carters, particularly Jay Sr., have made to advance the art of wind turbine design. "I'm fascinated by the Wind Eagle," confessed Osborn, the former Kenetech president. "It represents the kind of thinking that will take the technology to the next step. The beauty is, it is so simple. All you need is a pickup and a post-hole digger. The turbines can be relocated, which could become a major plus."

Peter Munser, the last CEO of Carter Wind Turbines shakes his head and lets loose the deepest of sighs when you bring up either Carter's name. "There is nothing left," he said of the Carter business. Munser was in charge after Junior threw a temper tantrum and filed a lawsuit against his dad, claiming he had the property rights to their original machine. There was an "investor dispute" and the board of directors moved the company to Europe, where it was sold to interests in India. "Their turbines were radical designs. Like a Studebaker, they were a most admired car but people were afraid of them. The problem with the original turbine was that when it started up, it whipped back and forth like a palm tree. That's when the blades can hit the towers. The bigger you go—like 300 kW—the larger the problem," said Munser. He told me the turbine had "design flaws that were never corrected" in part because Carter Sr. is very possessive about his technology. "God's gift to the design world, that's what he thought of himself," said Munser. I asked him what the bottom-line lesson was with the Carters. "You may have a great design and a great engineer, but they don't necessarily make good businessmen. You talk to both of them. They both want to be God; they want to be everything."

Charren, the money and the brains behind Kenetech, is still bullish on wind technology. I completed our interview at the Charles Hotel, in Cambridge, by asking him what he thought about the future of wind power. He abandoned his previously measured demeanor and indulged me with the following math: "What will drive the wind power market will be the next fuel crisis. It may be in three years, it may be in five, but definitely within the next ten years. Why will there by an energy crisis? During the time we've been talking here—it's been about an hour and a half—15,000 people have been born. That equates to 250,000 people coming into this world every day. In ten years' time, that is 2 billion people. Our current consumption rates, without population increases factored in, will use up all 200 billion barrels of oil in the Caspian Sea within a couple of years. At some point, society will turn to renewable energy. Wind power is the most concentrated form of energy, it is the most efficient conversion of renewable energy, and, I bet you, it is going to win."

◆

Two trends will determine whether wind farmers truly deliver on the promise of a sustainable renewable energy future. The first is the deregulation of electricity markets that is sweeping the nation and the globe. In the short run, this trend has hurt the wind farming industry, as many utility monopolies abandoned wind projects that were on the drawing board, their rationale being that wind power was too expensive. In a competitive market, cheap power was the goal. Overpriced renewable energy projects would be a burden in this brave new world of choice. Over the long term, however, restructuring electricity markets could open up new opportunities for consumers to vote for cleaner power sources, in much the same way as the Sacramento Municipal Utility District's customers voted for renewable energy sources after the closure of the Rancho Seco nuclear power plant.

Wind power has emerged as the clear favorite among those consumers who are picking green power in California, Colorado, Texas, and Pennsylvania. Since wind farms can be put up one turbine at a time, the technology can accommodate the incremental nature of building new clean power plants based on voluntary individual customer sign-ups.

The other major trend driving the market for wind power in the future is concern over global climate change. In Europe, where wind power capacity is projected to reach 40,000 MW by 2010, countries have established specific targets for reducing the carbon dioxide linked to rising temperatures over the past few decades. Wind power has become the technology of choice to reduce CO_2 emissions. The United States lags behind Europe in implementing policies that would enable us to meet any CO_2 reduction targets. The task ahead is truly mind-boggling. For California to reduce CO_2 emissions to 1990 levels in its electricity sector alone would require that some 30,000 MW of new wind capacity be added by the year 2007. That is, California alone would have to install more than 4,000 MW of wind power each year for seven consecutive years! That's just in one state! The amount of wind power added in California during the first year of deregulation directly attributed to customer demand? Less than 5 MW!

Though most of the recent growth in wind power sales has been overseas—shrinking California's share of the global market from 90 percent in the early '80s to less than 10 percent today—the United States set a record for installations within a 12-month period ending June 1999. During this time over 1,000 MW of new wind power capacity was added, representing $1 billion in investment. No other country has matched this pace of wind power development in a single year. In recognition of this success and America's great and still largely untapped wind resource, DOE launched its strategic "Windpowering America" initiative to derive 5 percent of our electricity from wind power by the year 2020—which equates to 80,000 MW of new wind power capacity. That, too, equates to 4,000 MW per year—four times the recent one-year record—every year for the next 20.

While Denmark has long been considered the world leader in producing and consuming wind power (roughly 8 percent of the nation's electricity is supplied by over 1,350 MW of wind power), Germany has emerged as the new world leader, with more than 3,000 MW installed since 1991. The U.S. total spanning almost two decades of development now stands at about 2,500 MW.

Today, total global wind power capacity accounts for over 15,000 MW. This may seem like a large number, but it represents just over 0.1 percent of global consumption. Wind power has made great strides since the days of the California wind rush, but the road ahead is long.

Where there is wind, there is turbulence, friction, and ultimately, conflicts. On the physical plane, winds are all about chaos and particles of matter so minute they escape detection, yet they possess incredible amounts of energy. Chaos also dominates humanity's efforts to harvest the power of the wind, particularly in California, where the Altamont Pass serves to underscore the fine line one must tread in order to reach the promised land of a renewable energy future.

The sheer fickleness of the wind sometimes makes me think that the technology may never be perfected. Maybe nature is just something that cannot be completely tamed and counted upon in a business like supplying reliable electricity. The story of the wind farmers forces us to look at the delicate balancing act required to foster economic development and take advantage of renewable energy sources while still respecting the limits of this earth and its surroundings. The wind farmers bring into focus a fundamental quandary: If we want to continue to power the needs of our growing populations and their gadgets and toys, we have no choice but the wind.

Epilogue

Winds of Change

> Of all natural forces, the wind has always been the most difficult to grasp. It touches us, moves us, but we cannot touch back. It was our first experience of the ineffable. Something of the indescribable power, too remote to be seen, but near enough to be sensed in a very intimate, a very personal, way.
>
> —Lyall Watson

William Heronemus, Woody Stoddard, and Randy Tinkerman rarely show up at any wind power conferences. Yet their stories say the most about the see-saw history of this technology so connected to the landscape and all of its physical, cultural, and financial dimensions. Most of those involved in the wind farming business don't even know their own history, don't realize what a long, strange trip it's been. The flocks of people now migrating to the wind power industry (the wind power conference held in Palm Springs in May 2000 set an all-time attendance record) will write just the latest chapter in this saga.

Hard to believe that Heronemus was once bandied about as the vice-presidential running mate of Barry Commoner when Commoner made an independent bid for the presidency in 1980. Heronemus's predictions about emerging clean power technologies such as wind power and even fuel cells way back in the '70s have largely come true. But no one seems to remember.

The American Wind Energy Association celebrated its twenty-fifth anniversary in June 1999 in Vermont and finally gave Heronemus some well-deserved recognition with a lifetime achievement award. The Smith-Putnam machine installed at Grandpa's Knob proved that the kinetic energy of the wind could be captured and converted into electricity and then fed into the monopoly utility's electric grid. Professor Heronemus would never have started up his college class had he not discovered the story of the Smith-Putnam machine once he became disillusioned with the false promises of nuclear power and left the Navy. Nor would that wind engineering class have birthed U.S. Windpower/Kenetech and the concept of corporate wind farming.

Heronemus's lifetime achievement award was a fitting tribute to a true visionary who inspired so many of the people who would become the backbone of the industry—the engineers and folks who worked in the labs and in the field, who dealt with the everyday challenges associated with wind farming. It was Heronemus who first talked about harvesting the tremendous winds blowing offshore in the early 1970s. He was also the first to proclaim that wind was a vast fuel, particularly abundant in the Great Plains (the focus of today's wind power renaissance), and that it could easily forestall the need to split any more atoms to generate electricity.

Now, over two decades later, the offshore wind power market is booming in Europe, and the visions Heronemus had about our energy future ring truer than ever. Many of his ideas still seem wild, almost science fiction-like in their grandeur. Place gigantic wind ships hundreds of miles offshore to generate electricity that would then be converted into liquid hydrogen for use in fuel cells and other devices to power everything from cars to companies? It does sound too good to be true. But society needs individuals like Heronemus, people who think big and who can attach numbers and formulas and cost figures to the kind of grand schemes probably necessary to solve such huge problems as global climate change.

While Ulrich Hutter, his long-time rival who passed away a decade ago, is now celebrated in Germany as a prophet, Heronemus is a virtual unknown in the United States. He hasn't received a dime from any government agency since 1980. The wind engineering program he started still exists at the University of Massachusetts, but it is a shadow of its former self. Before the AWEA conference banquet, Heronemus was asked to join the organization. "They said they had no record of my membership," he told me with a wry, conspiratorial smile before we sat down to hear his speech. "I politely refused. I was hoping they didn't try to charge me for dinner," he snickered. Asked to be the founding father of AWEA in the mid-'70s, Heronemus turned the offer down because he perceived the parties involved to be a bunch of shady characters who were more interested in making money than in generating electricity. He didn't have any interest in serving as a figurehead for that kind of organization.

Good former college professor that he is, Heronemus prepared a 45-minute lecture for his AWEA award acceptance speech. They let him speak for only five minutes, so he cut to the chase. The simple laws of physics governing the conversion of matter into heat reveal the folly of continued reliance on nuclear and fossil fuels, he said. The effects of our power systems were now being felt in the rising temperatures associated with global climate change. "Those who accept global warming as a problem know that there is an absolute requirement for the earth to remain in thermal balance within our solar system. . . . There is only one ultimate solution to the global warming problem— total reliance upon solar energy . . . and the most productive of all solar energy processes is the wind energy process. . . ." Wind power needs to be developed at a steady and appropriate pace, "and our free market capitalistic system that we hold so dear will not do the job. There is need for massive governmental interference. If we wait for the private sector to reduce the greenhouse gases linked to our fossil fuel use, it will be too late." Heronemus's remarks seemed to be creating a bit of stir in the audience; many of its members were used to bowing before the altar of competitive power markets as the electricity supply became deregulated. "It was Governor Jerry 'Moonbeam' Brown who got it right," he proclaimed, calling Brown the only brave politician in America when it comes to promoting a sustainable energy policy. "He said that wind power was going to happen and even though electric utilities tried to undermine him, and he was spat upon by many, Brown took a risk and got this industry off the ground. We could lead the world in the development of wind power, and this would be a project worthy of the United States. I sincerely believe that. We should stop spending our money on wars and on missiles and all of the other crap we are doing today," he added before his time was up and the crowd gave him a roaring standing ovation.

The next day I was at the kitchen table in the Heronemus house in Amherst. Woody Stoddard showed up, too, his long curls falling out from under a beat-up baseball cap. "I loved your speech last night," he told Heronemus as he looked out the window at the chickadees picking at the seeds in the bird feeder. "Nobody could have said it better. What Heronemus said last night in five minutes was the same message he made when this whole wind power thing started."

As if on cue, Heronemus interjected: "We could set up wind ships offshore just south of the Aleutian Islands, or the Chilean or Argentina shelves, places where the winds really roar. We can create liquid ammonia by relying upon wind power to drive the electrolysis process to provide hydrogen. This liquid ammonia could be transported by ship, barge, or even pipeline. Under this plan, for which I am currently seeking funding, we could kill two birds with one stone. We could greatly reduce CO_2 emissions that are contributing to global climate change by displacing natural gas with renewable wind energy, and we could help feed the world."

"We could also use a similar process to make Portland cement," Stoddard chimed in. "That's right," said Heronemus. "The process of turning stone into Portland cement involves an awful lot of seagoing transport. That stone could be crushed with renewable energy from these same wind ships. That's another project worthy of the World Bank."

"I have to admit I'm really inspired," Stoddard remarked. "I feel like I've come full circle after the AWEA banquet last night. "There are so many places where we could put wind turbines—it could be huge. There is so much fucking land where we could put wind machines, it's not even funny. Why do you think the Great Plains is depopulated? There's too much damn wind!" He went on to say that this is the ideal place to put huge arrays of large wind machines because, over several years, turbines would be on-line two to three times as often as Kenetech's machines in the Altamont Pass. "We are talking [about wind farms operating] 60 to 70 percent [of the time]."

Stoddard suddenly said he had to leave. Heronemus explained, "Woody and I have been talking about starting up our wind power company, getting some of the original students—like Ted Van Duzen—to join in. We wanted to make ocean wind systems a reality. I know exactly how to do it, and I know who my partners should be. But Woody is really struggling right now. You know, he now teaches at one of the worst high schools around. But he loves the challenge. His mom also has Alzheimer's and he feels he has to look after her. Enron wanted to hire him full-time, but he doesn't want to live in Tehachapi. On top of it all, all along I had been saving a sum [to start a wind power company], but just recently invested in some real estate. We have the guts, but now we still don't have the money."

There was a long silence. I could tell Heronemus was a little embarrassed by his recent investment decision, but he assured me he hadn't given up on the idea of obtaining some public or private funding source to support any one of his wind power proposals. It was indeed a bittersweet moment for both Heronemus and Stoddard. These early pioneers, the two who had provided the intellectual seed for what would become the largest wind farming company in the world, were still fantasizing about ways to create a wind power company with smarts, savvy, and a sense of camaraderie that lived up to the high ideals they both cherished.

◆

A year later, I ran into Randy Tinkerman again. He wasn't doing too well. "You might find me under a bridge," he sighed, wondering aloud how he would make his next rent payment. I took him to lunch and quickly discovered he really likes fine food and wine as he ran up my tab. Tinkerman was struggling because all of his straight talk over the years had alienated the powers that be, severely limiting his future work in wind farming.

He did finally find work at the San Francisco Dry Docks that lie near his Potrero Hill apartment. Believe it or not, Tinkerman actually worked on a ship that Heronemus used to serve on decades ago. By the time we spoke on the phone next, however, he had already quit the job. But he was very excited, just the same. "I ended up working on an oil tanker. When I was instructed to clean the crude from the actual holds, I gave it my best shot, but I began to feel ill.

You wouldn't believe what it is like to be inside of one of these huge tanks, covered with crude or tar. The fumes were getting to me so I finally told my supervisor, "I've spent my entire adult life fighting for renewable energy to replace fossil fuels. Down there is poison—I ain't going down there. He said 'Fine, go home.'"

As Tinkerman walked back to his apartment, on a whim he phoned Stoddard, who had called him recently for advice on wind business.

"So, I guess you saw my e-mail," Stoddard said.

"What e-mail?" Tinkerman asked.

"You mean you just called me out of the blue?"

"Yes, I felt compelled to call you."

Heronemus and Stoddard had decided to start up a new wind company, and Tinkerman was again helping out. They had decided that very morning and Stoddard had e-mailed Tinkerman to let him know about the decision. They would try, one last time, to build an American wind power company that would live up to the ideals they first embraced in the 1970s. "You can't tell anyone yet," Tinkerman said. "They are talking to big-league investors, folks I can't mention."

Appendix

Questions of Balance

Black-throated wind, whisper with sin
And speak of a life that passes like dew.
—John Barlow

It is 11 o'clock on a June morning in the Altamont Pass. The wind is calling on a dry bronze ridge that sits at the crest of this Range of Good Winds that became the obsession that then became this book.

I am not alone, though. Dr. Met Kreithen is aiming his video camera and laser beam at homing pigeons numbers 206 and 1114 as they attempt to fly into the 20 mph gusts of wind and back to their loft, which is obstructed by whirling propellers that populate the view. They approach the string of wildly spinning machines from the east and fly westward, shoot through the large, bright white 100-foot blades of Kenetech's imposing 33M-VS turbines. For some odd reason, though, both birds quickly come back toward us, through a spot where a bladeless machine provides a convenient opening in the string of turbines. They then shoot back through the spinning blades at yet another part of the string and then finally they plunge back toward us again and ultimately continue on toward their loft, which sits near the Tesla substation to the east.

Kreithen has studied pigeons for 24 years. As he enters the data from this latest flight into his Apple laptop, he proclaims these experiments "make the invisible visible." He and other members of his research team had already conducted 7,000 of these pigeon turbine interactions, he explained. Three strikes, two fatalities have occurred. It would take 31 years, on average, to witness an eagle colliding with a turbine in the wild, he told me. The flights I witnessed helped in the development of a design that was later painted on turbine blades. "Eyes as sharp as an eagle is a myth," stated Kreithen. Though birds can see ultraviolet light, they do not perceive contrast as well as humans do, and the painted design took into account current research into the nature of their vision. The plan was to test patterns on pigeons in the Altamont. Then, raptors, with monitoring devices attached, were supposed to be released and tracked as they navigated through and, it was hoped, dodged the turbines.

Some bird watchers, myself included, have tempered their judgments of Kenetech because of the company's funding of Kreithen. Research underwritten by Kenetech had some convincing solutions to the avian mortality dilemma. Results from one study suggested that larger turbines sitting atop tubular towers with slow-moving rotor blades and variable-speed controls kill far fewer birds. According to one estimate, if Kenetech replaced its entire fleet of 56-100s with its new model, which included all of these features, the probability of bird collisions would drop by 80 percent. Hans Peters, a local biologist who has helped focus attention on the raptor kill problem at the Altamont, praised the company. "I'm no fan of wind turbines, but I'll say this of Kenetech: they're trying." Peters went on to note that the firm has already installed "perch guards" on a number of the wind turbines to reduce roosting, which Peters believes is a major cause of raptor mortality. Peters claims these guards have reduced roosting by almost 50 percent.

The state study that revealed the disturbing levels of eagle kills in the Altamont Pass led to the creation of a special task force of the nation's premier avian scientists that included Kriethen. The effort was headed by Dr. Tom Cade, professor emeritus of ornithology at Cornell University. Cade, a bald, unassuming man who looks a bit like President Eisenhower, has been credited with bringing back the Peregrine Falcon from the brink of extinction due to widespread use of DDT. His new job, in the twilight of his career, was far more difficult.

Cade was hired by Kenetech to work with the country's top scientists to figure out why birds collide with wind turbines. This is no easy task, because only a handful of strikes had been witnessed in the first 20 years of wind farm operations. The first involved a huge, solo turbine—the one PG&E put up with DOE funds—several years back. A hawk of unknown species was playing, dodging in and out of the range of the blades, when it was finally struck. Members of a task force on avian mortality saw a kestrel run into the nacelle of a nonoperating turbine. (I was later told that kestrels are notoriously dumb birds.) Another fatality was witnessed by a windsmith: A red-tailed hawk came up to a spinning turbine, spreading its wings as if to perch, and was hit.

Perching is a bigger part of the problem than originally thought, Cade remarked. "Raptors frequently sit on towers when they are not operating, using these structures as perches on quiet days. When turbines become active, they may be startled into flight." Yet another theory, Cade noted, is that "birds get into the habit of using particular towers as perches. Sometimes they may not be aware the turbine is operating. They fly up to their typical perch and get struck." Cade offered his perspective. "For some, one individual bird getting killed is a severe problem. My own personal feeling is that [the level of mortality] is not significant." He observed that a wind farm area the size of the Altamont will never likely be repeated. "Therefore, new wind farms will be spreading the impact over a larger geographical area. Even if the same bird strike per turbine ratio as Altamont is assumed (fewer than three strikes per 100 turbines annually), there will be less significant impacts on populations," he said.

"The long-term future of the eagles at the Altamont is dependent upon a host of other factors besides wind turbines," Cade continued, noting that the turbines preserve the big ranches that dominate the landscape. "If this area is subdivided, nest sites will be gone—and there goes the eagle population." A federal Fish and Wildlife Service study of avian mortality in general provides some context to Cade's observations. It found, for instance, that our highways kill 57 million birds every year. While the leading cause of bird death is hunters, other sources of mortality include buildings with plate glass facades (97 million), pollution and poisoning (3,815,000), and tall buildings (1,250,000). Even domestic cats are part of the problem; felines devour millions and millions of songbirds in America every year! And the Exxon Valdez oil spill—a sad reminder of costs associated with our current addiction to fossil fuels—killed at least 300,000 birds, including more than a few bald eagles.

Grainger Hunt, an ecologist with the Predatory Bird Research Group of the University of California–Santa Cruz, has the most direct pulse on whether populations of eagles at the Altamont are in jeopardy. Since January 1994, 179 golden eagles have been radio-tagged and monitored by Hunt. Of the 61 recorded deaths of radio-tagged eagles, 23, or 38 percent, were caused by wind turbine blade strikes. Additional deaths were attributed to blade strikes but went unrecorded because radio transmitters were destroyed in the collision. Hunt said, "Adult eagle breeders are the most important" in terms of population dynamics, and these birds are "not too vulnerable" since they stay within their nesting territories, trees that lie on the outskirts of the wind farming region. "They rarely enter the wind farm area," he said. At least 70 golden eagle breeding territories are located in the Altamont Pass region, and all of these sites were fully occupied during the study.

The whole key to the long-term survival of the golden eagle population here is the balance between adult breeders and a population of "floaters," a term used to describe adults waiting in line for new nesting sites once existing breeders die. So-called "floaters" are critical. "They buffer breeding populations from significant loss," said Hunt. Since only 3 percent of the breeding pairs he monitored were subadults rather than mature floaters, there appears to be a healthy population dynamic at work. The critical question is this: "Does the death rate outstrip the birth rate?" So far, the evidence is inconclusive. If there is too much competition for breeding sites, the survival rate of nests diminishes. If there are not enough floaters, breeding sites may be vacant and the population suffers too. To date, there appears to be some balance to the population, though Hunt declared in a report to the National Renewable Energy Laboratory that the overall population of golden eagles is in slight decline.

According to Paul Kerlinger, a scientist and author specializing in bird behavior, it remains unclear whether Hunt's evidence of a small decline is due to the wind farms or human land use patterns and development in the region. Hunt doesn't buy the argument that wind farms have increased the golden eagle population. Kerlinger thinks other dynamics are at play. Kerlinger pointed out that since the mid-'90s, when Kenetech's bird monitoring program began, "a new reservoir destroyed foraging and nesting habitat within the wind resource area; the City of Livermore has pushed its boundaries up against the wind farms; and it is beginning to look like the oak trees that exist outside the wind resource area are not coming back due to cattle grazing. These are the oaks golden eagles nest in. As the trees go, so go the eagles."

It is not just the wind industry, and scientists like Cade, who think the Altamont wind farms may be preserving local raptors like the golden eagle, Kerlinger remarked. Even the FWS acknowledges the negative impact of the ever-expanding San Francisco Bay Area urban populations on this uniquely positioned golden eagle population. Ken Sanchez, a FWS senior biologist based in Sacramento, used the word "hammered' to describe the impact that burgeoning growth nearby has had on what he described as the "highest-density golden eagle nesting site" known. "I agree that wind farms are better than subdivisions," he said. As is the case with other complex environmental issues, regulators act only where they have distinct authority do to so. FWS has virtually no leverage on these other contributing factors, so they focus their enforcement activities on the wind farmers.

"This was probably the worst place in the world for a wind farm," Sanchez stated, despite his acknowledgment of the host of other contributing factors that will determine the long-term viability of the local golden eagle population. "We are still involved in an ongoing law enforcement action against these wind companies. The federal Department of Justice does not seem willing to prosecute, but that certainly isn't because of a lack of evidence." Sanchez pointed out that the golden eagle carcass total has surpassed 100 birds frozen in storage. He added that FWS is thinking of dropping criminal charges and instead pursuing civil remedies.

The Bald Eagle Protection Act extends to golden eagles and, if taken literally, forbids a single kill—even by accident—of a single eagle. Apparently there are more than a few FWS employees who think the wind farmers of the Altamont all belong in jail.

However, in 1998 Dick Curry, the former director of Kenetech's avian task force, and Kerlinger compiled results from years of field testing in the Altamont Pass. They say the results show that the wind industry can implement programs that would greatly reduce the tally of dead birds. This incredibly thorough analysis has been instrumental in persuading local county officials to approve the removing of what have been identified as the Altamont's "killer" wind turbines. Perhaps the biggest lesson learned from their research is the importance of topography and placement. Only 459 (or 13 percent) of Kenetech's 3,400 small 56-100 turbines have been implicated in bird strikes. Furthermore, only 16 turbines (4.8 percent) have killed one, two, or three eagles over a nine-year period. Typically these are at the end, or second from the end, of a string of wind turbines located on a steep hillside. Although about a third of the company's turbines fit this description, they account for 46 percent of golden eagle deaths. And although more than half the eagle and red-tailed hawk kills occurred at mid-string turbines, those located in dips and valleys and spaced irregularly stood out. All told, topography and placement explain 68 percent of eagle kills and 60 percent of red-tailed hawk kills.

It was these persuasive data presented on behalf of Green Ridge Services, which purchased part of the bankrupt Kenetech assets, that convinced county officials to approve a repowering plan to replace 1,300 of Kenetech's 56-100s, focusing on removing as many of the identified "killer" turbines as possible, and instead installing only 200 of the 700 kW Danish Micon wind turbines. Kerlinger has calculated that this change-out would reduce the probability of golden eagle strikes in the retrofit area by 45 percent while providing the same amount of electricity. "The Altamont Pass is an anomaly," Kerlinger said. Recent California Energy Commission reports support this claim. Both Tehachapi and San Gorgonio have far lower levels of avian mortality. Indeed, initial monitoring efforts identified not a single eagle carcass in either wind farming area. The biggest problem in Tehachapi is great horned owls, whose nocturnal hunting habits are apparently disrupted occasionally by a spinning wind turbine; almost half of the small number of raptor deaths were of these owls. While there are few raptors in Tehachapi—I spotted only one during a five-hour wild flower hike through the wind farming region—as a group they still had the greatest risk of strikes. All told, 77 dead birds were found in Tehachapi within a one-year period.

San Gorgonio has a much larger population of birds than Tehachapi. Yet its levels of bird mortality are the lowest among California's major wind farming areas. Only 29 birds were found dead near wind turbines here, with rock doves being the species most frequently killed. A report issued by Southern California Edison claimed that 6,800 small song birds were slaughtered in San Gorgonio in a single migration season, and suggested that they were perhaps a result of the dense packing in of wind turbines at some sites. The CEC report found no evidence to back up this claim of annual massacres. Only one passerine was found among the dead birds.

The single largest factor in predicting raptor mortality rates is raptor population. The Altamont Pass has 10 to 20 times more raptors than Tehachapi and 200 to 300 as many as San Gorgonio.

Some biologists aren't too impressed with the work of Curry and Kerlinger and think the bird problem at the Altamont Pass is probably understated. "The data bases people like Kerlinger rely upon are junk," said Carl Thelander, a bird expert with BioResource Consultants. "All of the numbers he and the wind industry use rely upon dead birds reported by the windsmiths in the field. Well, we've had some trained biologists go out and find carcasses that the windsmiths missed." Thelander acknowledges that the Altamont is an anomaly, but is very disappointed in the wind industry's slow response to this touchy topic. He added, "One computer model shows no change in the current golden eagle population in the Altamont, the other shows a slight decline. But neither model is really valid. I think the problem could be solved with the application of adequate resources." Thelander suggested a goal of a 50 percent reduction in raptor mortality within five years. "I have no doubt that with the proper funding, we could reach that goal," he said. "This is not an insurmountable challenge."

Source Notes

Chapter 1

The material in this chapter was derived from in-person and phone interviews with Randy Tinkerman conducted sporadically from the summer of 1992 through March 2000; reflections gleaned from my own private as well as official tours through the Altamont Pass and many of its significant sites; and an article, "Wind Power," by Kerry Drager in the *California Journal* (April 1981), p. 152.

Chapter 2

Material about the names different cultures have bestowed upon wind and about the many myths attached to its power was collected from:

Jan DeBlieu, *Wind: How the Flow of Air Shaped Life, Myth and Land* (New York: Houghton Mifflin Company, 1998), pp. 175–182.

Jerry Dennis, *It's Raining Frogs and Fishes: Four Seasons of Natural Phenomena and Oddities of the Sky* (New York: HarperCollins, 1992), pp. 7–16.

Voltra Torrey, *Wind-Catchers: American Windmills of Yesterday and Tomorrow* (Brattleboro, VT: Stephen Greene Press, 1976), pp. 16, 17.

Lloyd Watson, *Heaven's Breath: A Natural History of the Wind* (New York: William Morrow & Co., 1984), pp. 7, 327.

A. B. C. Whipple, *Storm* (Alexandria, VA: Time-Life Books, 1982), p. 43.

Descriptions of Navajo and other indigenous peoples' beliefs about the wind were drawn from an interview with Luci Tapahonso, a Navajo poet and writer, at the "Art of the Wild" conference in Squaw Valley, July 1999; Gary Witherspoon, *Language and Art in the Navajo Universe* (Ann Arbor: University of Michigan Press, 1977), pp. 30–41; Timothy Eagan, *Lasso the Wind: Away to the New West* (New York: Alfred A. Knopf, 1998), pp. 64–68; Jan DeBlieu's *Wind*, pp. 28–35; and from the liner notes for Dik Darnell's compact disc, *Voice of the Four Winds.*

An interview with Debby Jolly, an acupuncturist, in March 1999 provided information about views on seasons and emotions found in Chinese medicine.

Information about early scientific inquires into the physics of wind, weather systems, and jet streams was derived from:

Elmar Reiter, *Jet Streams: How Do They Affect Our Weather?* (Garden City, NY: Doubleday & Co., 1967), pp.1–11.

A .B. C. Whipple's *Storm*, p. 54.

Dennis's *It's Raining Frogs and Fishes*, p. 19–22, 133–139.

Nicholas P. Cheremisinoff, *Fundamentals of Wind Energy* (Ann Arbor, MI: Ann Arbor Science, 1979), pp. 1–2.

Antoine de Saint-Exupery, *Wind, Sand & Stars* (London: Harcourt Brace & Co., 1939; translated by Lewis Galantiere, 1969), p. 55.

DeBlieu's *Wind*, pp. 144–148.

Watson's *Heaven's Breath*, pp. 55–56, 155.

Descriptions of the peculiarities of the Altamont Pass wind resource were derived from interviews with Tinkerman, several articles published in assorted newspapers and magazines in the early '80s, and discussions with California Energy Commission staff.

Chapter 3

Information about wind witches and early sales of wind rights was collected from:

Watson's *Heaven's Breath,* pp. 118–119.

DeBlieu's *Wind,* pp. 33–34

Dennis's *It's Raining Frogs and Fishes,* p. 8

These books provided information on the early development of Dutch and English windmills:

Robert Righter, *Wind Energy in America* (Norman: University of Oklahoma Press, 1996), pp. 7–16.

Paul Gipe, *Wind Energy Comes of Age* (New York: John Wiley & Sons, 1995), pp. 118–123.

Torrey's *Wind-Catchers,* pp. 50–60.

Frederick Stokhuyzen, *The Dutch Windmill* (Bussum, Holland: C.A.J. van Dishoeck, 1962), pp. 87–88.

Edward Kealey, *Harvesting the Air: Windmill Pioneers in Twelfth-Century England* (Berkeley: University of California Press, 1987), pp. 50, 57, 69–70, 107–108, 132–153.

Lynn White, *Medieval Religion and Technology* (Berkeley: University of California Press, 1978), p. 22.

The information about Stockton, "City of Windmills," was based largely on interviews and leads provided by Jim Williams, director of the California History Center & Foundation, and Chapter 4, "Wind, Tide and Sun," from his manuscript, "Energy and the Making of Modern California," (Cuptertino, CA: De Anza College, 1993), pp. 79–89; and interviews with representatives of, and documents provided by, the San Joaquin Historical Society, including an article by Edrie Bastian, "Stockton, the City of Windmills," the cover story in *The San Joaquin Historian,* Spring 1987.

Details about the evolution of windmills outside of California were derived in part from Paul Vosburgh, *Commercial Applications of Wind Power* (New York: Van Nostrand Reinhold Co., 1983), pp. 205–208.

Chapter 4

Most of the information about Nikola Tesla was culled from Margaret Cheney, *Tesla: Man out of Time* (New York: Dell, 1983), including his endorsement of wind power (p. 162), his late-night antics with Mark Twain (p. 3), his strained relationship with Thomas Edison (pp. 24–37), the battle over AC (pp. 38–50), and the notion Tesla may have been from another planet (pp. 81–82).

Information about Tesla's first patents (p. 18) and his inventions in the twilight of his career, such as the Death Ray (p. 247), were found in a book compiled by David Hatcher Childress but listed as "authored" by Tesla: *The Fantastic Inventions of Nikola Tesla* (Kemton, IL: Adventures Unlimited Press, 1993).

The contention that Edison did not really invent the light bulb and many other inventions attributed to him is addressed in Robert Friedel and Paul Israel, *Edison's Electric Light* (New Brunswick, NJ: Rutgers University Press, 1987), pp. xii–xiv, 224.

Cheney's *Tesla* was a source of much of the information about Edison, as was Paul Israel, *Edison: A Life of Invention* (New York: John Wiley & Sons, 1998), which provided facts about his Menlo Park laboratory (pp. 122–125, 191–201), his stubborn advocacy of DC (pp. 326–337), and questions about the exact nature of his contributions to inventions attributed solely to him (pp. 372–378).

Ed Smeloff and Peter Asmus, *Reinventing Electric Utilities: Competition, Citizen Action and Clean Power* (Washington, D.C.: Island Press, 1997) provided background on the emergence of electric utility monopolies at the turn of the past century and the subsequent evolution of the public power movement (pp. 9–13).

Joe Kaiser, *Electrical Power: Motors, Controls, Generators and Transformers* (Tinley Park, IL: Goodheart-Willcox, Co., 1998) provided information about the differences between AC and DC current (pp. 40–52).

The information about Charles Brush was largely derived from Righter's *Wind Energy in America,* pp. 42–58.

Information for the profile of Marcellus Jacobs came from Jon Naar, *The New Wind Power* (New York: Penguin Books, 1982), pp. 70–72; Paul Vosburgh's *Commercial Appli-*

cations of Wind Power, p. 210; and Frank Eldridge, *Wind Machines* (New York: Van Nostrand Reinhold Co., 1980), p. 24. I also interviewed Phil Leen, an employee of the Solar Depot of Sacramento, California, and former member of the American Wind Energy Association. Leen worked directly with Jacobs and is a hands-on expert on small wind turbines.

Chapter 5

The information on German gliders was found in Lloyd Watson's *Heaven's Breath,* p.134.

The information on Ulrich Hutter was compiled from:

Gipe's *Wind Energy Comes of Age,* pp. 77–80.

Vosburgh's *Commercial Applications of Wind Power,* p. 222.

Eldridge's *Wind Machines,* which also provided the basis for the history of Russian, French, German, and Danish advances in wind power (pp. 25–73).

Hutter was also discussed quite extensively during an interview in November 1998 with William Heronemus.

Palmer Cosslett Putnam, *Power from the Wind* (New York: Van Nostrand Reinhold Co., 1948), pp. 98–108, provided a historical overview of the evolution of wind turbines in Europe and Russia up until WWII.

Gipe's *Wind Energy Comes of Age* also provided key facts and figures regarding the evolution of the Danish wind power market (pp. 79–81) and Darrieus's vertical-axis machines (pp. 172–174).

A draft paper by Peter Karnoe, "Approaches to Innovation in Modern Wind Technology" (Copenhagen: Institute for Organization and Industrial Sociology; Stanford, CA: Stanford University Department of Economics, December 1992), also contributed to the section concerning the evolution of wind technology in Denmark during WWII, particularly pp. 24–29 (Brush and the Danes) and pp. 39–43 (Gedser and MOD-2).

The section of the chapter devoted to Palmer Putnam was derived from:

Putnam's *Power from the Wind,* pp. v–vi, xi–xii, 1–14.

Torrey's *Wind-Catchers,* pp. 130–140.

Grant Voaden, "The Smith-Putnam Wind Turbine: A Step Forward in Aero-Electric Power Research," *Turbine Topics* (a publication of the S. Morgan Smith Co.), Vol. 1, No. 3, June 1943.

Assorted articles included "Big Windmill," *Fortune,* November 1941.

Chapter 6

Most of the information about Heronemus was gathered during two in-person recorded interviews with him, in November 1998 and June 1999. Woody Stoddard participated in the second. Stoddard and Tinkerman also provided background on Heronemus and the key role he played in the early evolution of the modern wind power industry. I recorded an interview with Stoddard in November 1998, and he handed over a huge box of original documents that served as the basis for much of the early history of U.S. Windpower and Stoddard's subsequent falling out with the company.

Two books also provided quotes and some historical details:

Torrey's *Wind-Catchers,* pp. 166–173.

Righter's *Wind Energy in America,* pp. 155–156.

Much of the material about early DOE work on wind turbines and the role of Divone in the Federal Wind Energy Program was drawn from Adam Serchuk's excellent Ph.D. dissertation, *Federal Giants and Wind Energy Entrepreneurs: Utility-Scale Windpower in America,* 1970–1990 (Blacksburg, VA: Virginia Polytechnic Institute, January 27, 1995).

Details about the 1973 wind power workshop (p. 68), the General Electric and Lockheed studies (pp. 82–85), the quote from Hans Meyer (p.109), the dimensions of the early DOE machines (p. 90) were all found in Serchuk's thorough document. In addition, I interviewed Divone in his Washington, D.C., office in November 1998. His "Evolution of Modern Wind Turbines," Chapter Three in *Wind Turbine Technology: Fundamental Concepts of Wind Turbine Engineering,* Dick Spera, editor (New York: ASME Press, 1994), was the source of information about the Gedser mill (p. 74), Percy Thomas's dual-rotor proposal (p. 78), Hutter (p. 79), and the early start-up of the DOE program (pp. 80–84). Information about DOE expenditures was also found in Ray Reece, *The Sun Betrayed* (Boston: South End Press, 1979), p. 89.

Information about PG&E's MOD-2 was derived from interviews with Tinkerman in the summer of 1992 and with Phil Leen, who toured the site with Marcellus Jacobs in the fall of 1994, and from material found in Serchuk's *Federal Giants and Wind Energy Entrepreneurs,* pp. 185–190.

Divone's quote was taken from Serchuk's *Federal Giants and Wind Energy Entrepreneurs,* p. 140.

Chapter 7

Most of this chapter is based on interviews. I recorded interviews with Stanley Charren first in Cambridge, Massachusetts, in November 1998 and subsequently by phone in March 1999 to fill out the story of how he was persuaded to help finance a wind power company. I recorded interviews with William Heronemus and Woody Stoddard in Amherst on the same November trip and again with both in June 1999, also in Heronemus's residence, after the American Wind Energy Association conference in Burlington, Vermont. I made subsequent phone calls to both Stoddard and Heronemus during 1999 and in the spring of 2000.

Stoddard also lent me his extensive collection of original documents regarding the initial business plan of U.S. Windpower and numerous other memos, articles, and private placement memorandums.

Some of the early business details of U.S. Windpower were found in the following original documents: A January 31, 1980, letter to stockholders written by Herman Moore highlights the hiring of Herbert Weiss; a letter written on the same day by Stanley Charren highlights progress the company was making in raising $5.4 million on the "red book." A private placement memorandum written by Merrill Lynch, Pierce, Fenner & Smith for 2,400 units of limited partnership interests entitled "Windpower Partners 183-1" underscores U.S. Windpower's unique approach to wind power development (p. 21) and highlights the technical details of the wind turbine's operations (pp. 22–24).

Details about the Crotched Mountain wind farm installation, and quotes from Russell Wolfe, were found in Peter Fossel's "Harvesting The Wind," *Country Journal*, January 1983, pp. 49–53.

The information about Herman Moore's infamous "red book" was supplied in interviews with Stoddard and Charren and in descriptions in John Berger's *Charging Ahead: The Business of Renewable Energy and What It Means for America* (New York: Henry Holt & Co., 1997), pp. 146–148.

The anecdote about Herman Moore and Louis Manfredi was first told to me by Tinkerman in 1996 and was confirmed by Stoddard in November 1998. Some of the details were also revealed in Berger's book in the aforementioned pages.

Stoddard's lists of problems with the USW prototype machine and his expressed reservations about the direction in which the company was going appeared in a memo he wrote under the pseudonym "Storrow Woodward," on April 2, 1980.

Chapter 8

I first met Ty Cashman at the 1991 wind power conference held in Palm Springs. My most detailed talk with him about how he came up with the tax credit idea, and his relationship to the Carl Jung devotees of the Wind Harvest Company, occurred in the fall of that year. I've talked to him on the phone intermittently over the past eight years. I visited him for an in-person interview in January 2000, at which point he provided the details of how he became head of the American Wind Energy Association and how he helped convince the wind power industry to come to California.

Some of the details of Cashman's involvement in wind power prior to his stint in the Jerry Brown administration were found in Robert Righter's *Wind Energy in America,* pp. 165–166.

In the spring of 1999 I interviewed Robert Judd, former manager at the Office of Appropriate Technology, to find out more about Cashman's role and Jerry Brown's level of involvement with wind power policy.

I first interviewed Bob Thomas and George Wagner, by phone, in 1991. Subsequent phone discussions fleshed out Thomas's history. I then interviewed Wagner in person, first in November 1993 and again in October 1998, at Dipsea's Cafe, to get both updates on the current status of the company and more details regarding Wagner's persistent fundraising efforts.

Documents that provided additional information about Wind Harvest include:

Kelly Zito, "New Era Begins for Wind Power," *Marin Independent Journal* (Novato, CA), September 22, 1998, B1.

Ricardo Sandoval, "Riding Winds of Energy Change," *San Francisco Examiner,* August 31, 1992, B-1.

Wind Harvest Company, *Investor Summary* (prospectus), Point Reyes Station, California (no date).

Wind Harvest Company, *200,000 Shares of Common Stock* (prospectus), Point Reyes Station, April 22, 1998.

Chapter 9

A different presentation of much of this same material about the Altamont Pass appeared in my "Gone with the Wind," *Terra Nova: Nature & Culture* (Boston: MIT Press, Vol. 2, No. 2, Spring 1997), pp. 114–127.

The bulk of the original research for this chapter was collected on a hike through the Altamont Pass with Tinkerman in the summer of 1991 and a subsequent interview with Don Smith in June 1993, at a time when Smith was in the process of losing his long-time consulting gigs with Pacific Gas & Electric. Smith made copies of several documents, memos, and newspaper articles he had collected over the years that provided details about both the history of the Altamont Pass and the cooperative relationship between USW and PG&E.

Smith provided some of the information about local tribes in conversation. He also gave me copies of articles, one, "The Thin Green Line," from the *East Bay Express* (January 2, 1987), described Native American trading practices in the Altamont Pass region, and a copy of excerpts from a work whose title is so long I will abbreviate it: *Alameda County, including its Geology, Topography, Soil and Productions . . . and Incidents of Pioneer Life.* Published in Oakland by M.W. Wood in 1883, it was the source of most of the material about Native American nudity, sweats, and disease in the Altamont Pass.

I gathered my earliest material about Joaquin Murrieta from interviews with Bill Graham, the U.S. Windpower windsmith at the Montezuma Hills site, and his wife, Kathleen Gregorski, who used to hike in the Altamont as a kid and first told me about the local legends regarding Murrieta's buried treasure in the Altamont Pass.

At the Art of the Wild conference, in 1992, I interviewed Louis Owens, an author who integrates Native American themes into his work; he supplied me with critical leads for information about the Native American tribes that resided in the Altamont Pass region. Interestingly, he told me that the legend of Joaquin Murrieta was created by a half-Cherokee writer whose pen name was John Rollin Ridge. Ridge's book, *The Life and Adventures of Joaquin Murrieta, Celebrated California Bandit,* was originally published in 1854 in San Francisco. I found an edition that the University of Oklahoma Press published in 1955. The material in Ridge's book, particularly details about Murrieta's alleged activities near the Altamont Pass, was supplemented by Dianne C. Klyn, *Joaquin*

Murrieta in California (San Ramon, CA: The Publishing Place, 1989), pp. 11, 15, 29, 34.

C. Hart Merriam, *The Dawn of the World: Myths and Tales of the Miwok Indians of California* (Lincoln: University of Nebraska Press, 1993) supplied additional information about myths of local tribes, in particular golden eagles (pp. 17, 45–46, 163–167) and Mount Diablo and other raptors (pp. 67–90).

The report of the Altamont Speedway rock concert draws on a variety of sources, among them Bill Graham and Robert Greenfield, *Bill Graham Presents: My Life Inside Rock and Out* (New York: Doubleday, 1992), pp. 294 and 296; an e-mail exchange with Owsley Stanley in June 1999; and e-mail discussions among a few of the original participants on the thirtieth anniversary of the concert that were forwarded to me on December 9, 1999. Newspaper clips include Kevin Fagan, "Seeking Shelter from a Memory," *San Francisco Chronicle,* December 6, 1999, p. 1; Michael Mooney, "One Man's Dreams Were Buried with the Dead at the Altamont," *Oakland Tribune,* December 5, 1989, p. C-3; and an AP report "The Altamont Nightmare, 20 Years Ago Today," the *Oakland Tribune,* December 6, 1989.

Chapter 10

The opening sequence is based on an in-person recorded interview with Alvin Duskin in San Francisco in the summer of 1995, and was supplemented by information from John Berger's *Charging Ahead,* pp. 143–146. Duskin's description of the way the federal tax credits worked was augmented by Mike Hernacki's article "Windfalls," *Financial Planning,* November 1985, pp. 126–127. The information about the Crotched Mountain wind farm was found in "World's First SWECS Windfarm Built on Mountain Ridge in New Hampshire," *Wind Energy Report,* January, 1981, p. 1.

A memo written by Tom Hillesland of PG&E on March 13, 1981 documents the cooperation between USW and PG&E in the early days of the Altamont Pass installations. The quotes of Duskin and Moore regarding USW's first permit approvals appeared in Ann Bancroft, "Fair Breeze For Windmills," *San Francisco Chronicle,* April 16, 1981, p.B-1.

A recorded interview with John Eckland at his Lake Tahoe residence in July 1996 was the basis for the history of Fayette. Additional details about the beginning of Fayette were obtained from the files of Robert D. Kahn and Co., a public relations firm that represented many wind power companies. Kahn turned over his complete wind power files to me in 1993. Among that material was the 1984 Fayette annual report that described the evolution of Fayette (pp. 4–12). Janet Hopson, "They're Harvesting a New Cash Crop in California Hills," *Smithsonian,* November 1982, pp. 123–125, also provided background information about Fayette and the Altamont Pass.

The information about landowner wind royalties appeared in a story by John Miller "Windfall: Harvesting the Elements," *Oakland Tribune,* July 30, 1989, p. A-8.

Chapter 11

The story of Terry Mehrkam was first told me by my friend Phil Leen, who referred me to a few old-timers in the wind energy business to get more details. Eventually, I tracked down Mehrkam's brother, Doug, who sent me a package of clips and old promotional materials in early 1996. Follow-up interviews with Doug yielded more details about how well his brother's business was doing and about the design of the wind turbine.

Other published sources provided by Doug Mehrkam and used to develop this chapter include:

Nancy Ray, "State Safety Agency Orders Shutdown of County Wind Farms," *Los Angeles Times,* February 4, 1982, Part II, pp. J-1, J-8.

Michael Stoner, "The Wind Is Alive," *The County Magazine* (Berk County, PA), October 1976, pp. 18–19.

Other details about the weather the day of the accident and the initial reaction to Mehrkam's death came from my interview with Chuck Davenport, Mehrkam's financial partner, in his SeaWest corporate headquarters in San Diego in November 1998. Paul Gipe and Ty Cashman also provided their impressions of the event and what it meant to the wind industry over the phone in the summer of 1996.

Chapter 12

The information about Merrill Lynch's role in the evolution of California's "standard offer" power purchase contracts was derived from my interviews with Stoddard and Tinkerman and confirmed in an interview with Stanley Charren in November 1998.

The financial figures for the evolution of USW's business in the Altamont Pass were obtained from documents provided by Don Smith, including:

A PG&E memo apparently written by Steven Goschke, "Meeting Notes, U.S. Windpower Trip," February 5, 1985.

Pacific Gas & Electric, "U.S. Windpower, Inc. Proprietary and Confidential Memo," January 20, 1986.

Richard Lyons, "Demand for USW Notes Outstrips Supply," *Alternative Sources of Energy,* April 1987, pp. 28–29.

David Quarton, "Machine Features: U.S. Windpower, Inc.," *Windirections,* April 1986, pp. 22–23.

The information about how the standard offer pricing worked in the first few years of the wind energy boom came from an interview with Jan Hamrin in the summer of 1997.

The information about Fayette's cash flows in the first few years, through the end of the dual tax credits in 1985, was culled from an article by Jeff Greenwald, "Reaping the Wind: The Farmers of the Altamont Pass," Image Magazine, *San Francisco Examiner,* June 8, 1986, pp. 30–33, and an article by an unidentified reporter, "Altamont Leads the Wind Parade," *Tracy Press,* October 19, 1983.

The numbers regarding the performance of various wind turbines in the Altamont Pass were found in a report written by Don Smith, "The Wind Farms of the Altamont Pass Area," *Annual Reviews of Energy,* 1987, Vol. 12, pp. 168–170.

An interview with Barry Ziskin, former consultant with R. Lynette & Associates, provided the basic description of how Eckland ran his wind farming business. Because his father was an investor in Fayette, Ziskin was quite familiar with Eckland's business dealings.

The evolution of Fayette's wind turbine was described in Ros Davidson, "Wall Street Buying into Wind," *Windpower Monthly,* January 1988, p. 12.

The Ralph Koldinger quotes were found in "Fayette Owners Take on IRS," *Wind Letters,* the American Wind Energy Association newsletter, p. 4.

Chapter 13

The material regarding Pacific Wind & Solar, and the early history of wind farming in the Tehachapi Mountains, was derived from in-person interviews with Jan Hamrin in her San Francisco office in May 1996 and January 1999.

The history of the Tehachapi Pass region was obtained over the phone from a representative of the Kern County Historical Society and phone conversations with Paul Gipe in 1997.

Two in-person interviews with Jim Dehlsen, in December 1991 and the spring of 1994, at his ranch home in the Tehachapi Mountains supplied the basic information about the origins of Zond. Additional details were gathered in a series of phone interviews and work sessions in the Santa Barbara office of Dehlsen & Associates in the winter of 1997 and early spring of 1998.

Other primary published sources of information about Dehlsen's role in the history of Zond include:

John Farrell, "Reaping The Wind," *Westways,* November 1983, cover story.

Jessica Maxwell, "Wind in His Sales," *Western World,* July 1982, pp. 42–45, 51.

Jessica Maxwell, "Wind: The Next Multibillion-Dollar Business," *Esquire,* April 1983, p. 121.

Phone interviews were conducted with Ed Salter and Lloyd Herzinger in May 1992.

The information about Zond's alleged "lavish" profits was found in Ros Davidson, "Zond Salaries Reported Lavish," *Windpower Monthly,* August 1987, p. 12.

Chapter 14

The history of the San Gorgonio Pass was based on documents provided by the Riverside County Regional Park and Open Space District and was shaped by descriptions of the region found in Joan Didion's "The Santa Ana," which originally appeared in the

Saturday Evening Post and is included in *Natural State,* edited by Steven Gilbar (Berkeley: University of California Press, 1998), pp. 326–329. The information about Willie Boy was derived from Abraham Polanski's 1969 film *Tell Them Willie Boy is Here.*

The information about Dew Oliver was pulled from the following three press clips:

James Clebourne, "Reaping the Wind," *Palm Springs Villager,* April 1957.

Walter Ford, "Power from the Wind," *Desert Magazine,* June 1975, pp. 22–23.

Bob Pratte, "Wind Machines Nothing New to Those Who Saw Dew Oliver's Funnel," *Riverside Press Enterprise,* February 19, 1984, pp. B-1–B-2.

The following published clips served as the basis for the rest of the chapter:

"Bendix Turbine to be Dismantled," *The Desert Sun,* February 5, 1985, A-10.

James Throgmorton, *Community Energy Planning: Winds of Change from the San Gorgonio Pass* (Berkeley: University of California Press, 1987), pp. 361–364 (University of California Appropriate Technology Grants Program Project No. 82-222-4000).

Louis Sahagun, "Tilting at Windmills: Palm Springs Gallops into Court to Battle Energy Devices," *Los Angeles Times,* April 9, 1985, pp. 1, 6.

Fred Noble, "Windmills Generate Controversy in Palm Springs," *The Desert Sun,* July 1, 1989, p. E-1.

Ros Davidson, "Sonny Bono Takes His War on Windmills to Washington," *Windpower Monthly,* July 1989, p. 21.

Ken Ritter, "Bono Discovers New Tax Dollars Blowin' in Wind," *The Desert Sun,* August 12, 1990, p. A-10.

Ellen Paris, "Palm Springs and the Wind People," *Forbes,* June 3, 1985, pp. 170–171.

Phone interview with Gary Dodak, SeaWest employee, in July 1999 provided details of some of the more unusual wind turbine designs and installations in the San Gorgonio Pass.

The quotes from Chuck Davenport were collected in an interview conducted in November 1998.

Chapter 15

An interview with Bob Sherwin at his Vermont office in November 1998 provided the information about the early days of Enertech.

The bulk of the information about how the Carter turbine designs differed from the competition's and how the Carter Sr. and Carter Jr. designs differed from one another was obtained during an in-person interview with Carter Jr. at the 1993 AWEA conference in Denver. A March 1999 phone interview with Carter Sr. provided me with the other side of the story of this family feud.

Descriptions of some installations of Carter machines in the early '90s and some history of the company were reported in my article, "A Texas Firm Grows on Wind power," *In Business,* January/February 1994, p. 27.

Additional details about the history of the company and its finances were culled from two articles by Ros Davidson for *Windpower Monthly:* "Carter Design More Advanced Than Any Danish Concept," March 1987, pp. 12–13; "Carter Quietly Closes Doors to Business," February 1989, p. 8.

I interviewed Mike Bergey about Bergey Windpower several times by phone in early 1999 and in person in June 1999, at the AWEA conference in Burlington, Vermont. In February 1999 Bergey sent me copies of overheads he used in presentations about his company and the future of the small wind turbine market. Additional details about the start-up of Bergey Windpower were found in Ros Davidson's company profile, "Very Bullish About Prospects for the Future of Wind," *Windpower Monthly,* January 1988, pp. 14–16.

Chapter 16

The reporting of USW's problems with its machines was informed by my interview with Don Smith; his report, "The Wind Farms of the Altamont Pass Area," *Annual Reviews of Energy,* 1987, pp. 160–161, 168; the February 5, 1985, memo written by PG&E employee Steven Goschke; and a memo written by Smith on August 14, 1984.

The information about the bug build-up on wind turbine blades was found in a press release put out by the Sacramento-based Center for Energy Efficiency and Renewable Technologies (CEERT) in July 1997.

The information about how much federal and state tax credit revenues contributed to the development of California's wind industry was obtained from David Modisette, formerly on the staff of State Senator Gary Hart of Santa Barbara, author of the California wind power tax credit bill.

A study entitled *The Hidden Costs of Energy* (Washington DC: Center for Renewable Resources, October 1985), p. 2, provided figures comparing investments in renewables with those in conventional energy technologies. The comparison used in this chapter originally appeared in my article, "New Technologies Revitalize Wind Power," *California Journal,* August 1992, p. 414.

Much of the analysis of the success of the wind energy tax credits was found in Thomas Starrs, "Legislative Incentives and Energy Technologies," *Ecological Law Quarterly,* Vol. 15, 1988, pp. 103–118.

I conducted a phone interview with Andy Trenka in the summer of 1997.

The Cashman and Divone quotes were obtained in previously noted interviews.

The comparison figures between Danish and U.S. wind power programs were found in Peter Karnoe's "Approaches to Innovation in Modern Wind Technology," pp. 6–10, 43–46.

I conducted phone interviews with Jamie Chapman and William Chapman in December 1998.

Information about wind farm tax shelter abuses comes from:

William Canter, "Retired Couples Victims of Tax Shelter Bankruptcy," *Windpower Monthly,* August 1986, p. 4.

Anne Richards, "Ill Winds over Windmills," *California Journal,* February 1987, p. 36.

The Denis Hayes quote was taken from a transcript of a video documentary, "Power Struggle," that Jim Dehlsen loaned me.

Chapter 17

The rationale for the name change from USW to Kenetech was described in a June 7, 1988, memo from Stephan Jaspen, vice president of operations for Kenetech, to Don Smith, at PG&E.

The numbers used to illustrate the financial health of Kenetech come from two sources: Ros Davidson, "Prototype Running," *Windpower Monthly,* September 1989, p. 19; Thomas Lippman, "Future of Wind Power Gets a Lift," *The Washington Post,* November 17, 1991, p. h-1.

I conducted an in-person interview with Dale Osborn in June 1999 at the AWEA conference held in Vermont. Information obtained in the interview was supplemented by Ros Davidson, "Shaping Up For a New Future," *Windpower Monthly,* May 1989, p. 14.

I gathered information about the Sacramento Municipal Utility District during my ongoing reporting assignments for *California Energy Markets* and McGraw-Hill energy newsletters such as *Independent Power Report.*

Tinkerman published his retort to Osborn in "Wrong to Malign the Ponytailed Pioneers," *Windpower Monthly,* March 1990, p. 4.

The core research resource for my reporting on the issue of avian mortality was a study by Biosystems Analysis, Inc., *Wind Turbine Effects on Avian Activity, Habitat Use, and Mortality in Altamont Pass and Solano County Wind Resource Areas, 1989–1991* (Sacramento: California Energy Commission, March 1992), pp. ix–xii.

The account of the emergence of avian mortality as an issue surrounding wind power is largely reprinted from two of my previously published articles: "Who Owns The Wind?" *E Magazine,* May/June 1993, p. 18; "Hot Air, Hot Tempers and Cold Cash," *Amicus Journal,* Fall 1994, pp. 30–35.

Coverage of the proposed Zond wind farm at Gorman is based largely on a packet of information obtained in 1992 from Nancy Rader, who was working with Public Citizen at the time, and on interviews with Jim Dehlsen. Among the documents Rader provided were several Sierra Club memos.

The information about the illegal cement kiln comes from an article by Linda Dailey Paulson and me, "A Concrete Concern," *San Francisco Examiner,* April 23, 1995, p. B-3.

Other sources include two Ros Davidson *Windpower Monthly* stories: "Formidable Opponent Launches Dirty Tricks Campaign," December 1988, p. 9; "Zond Fights Propaganda with Profit Share Offer," January 1989, p. 6.

Chapter 18

This chapter is based on interviews conducted with windsmith Bill Graham in the spring of 1994 and a tour I took at Kenetech's Livermore facility in the fall of 1993. I interviewed John Opris several times by phone from the spring of 1994 through the spring of 2000. Some of this material appeared in my article, "Winds of Change," *Sacramento News & Review,* March 10, 1994, p. 21.

Chapter 19

I've talked to Robert Lynette on and off over the years. Most of the material in the story of FloWind was derived from a phone interview conducted in the spring of 1999.

The Wind Harvest material was derived from two interviews with George Wagner, referenced earlier.

He was hard to track down, but I finally interviewed John Kuhns in April 1999. His explanation of the details of the Fayette purchase was augmented by three articles by Ros Davidson in *Windpower Monthly:* "Swiss Lose out to Wall Street," March 1988, p. 14; "Wall Street Buying into Wind," January 1988, p. 12; "Fayette in Trouble Again," October 1989, p. 15.

Kuhns's slick maneuvers were described by a Tehachapi resident who has written a book or two about wind power.

The material regarding Kuhns's persuasive business practices in Chile was gathered during a consulting assignment for a client who must remain nameless.

Chapter 20

The fundamental source of information about the evolution in thinking behind Kenetech's new 33M-VS was "Excellent Forecast for Wind," *EPRI Journal,* June 1990, pp. 21–23.

Information about the *Enercon vs. Kenetech* lawsuit was provided in an Osborn interview and a subsequent phone interview with Mark Haller in September 1999. Additional details were found in an in-depth report by Henry Hermann, "The Wind Power Industry," for the New York investment banking firm WR Lazard, Laidlaw & Mead, November 22, 1994, pp. 26–27.

I obtained copies of Gerald Alderson's July 1, 1993, letter to Congressman Bill Baker and Baker's September 15, 1993, letter to Secretary of Energy Hazel O'Leary from the American Wind Energy Association.

The information about the success of Kenetech's original IPO and the funding it provided for developing the 33M-VS came from "Kenetech Pushes New Wind Technology with 6-Million Share Initial Offering," in the McGraw-Hill newsletter *Independent Power Report,* October 8, 1993, p. 8.

The roles of Merrill Lynch, Morgan Stanley & Co., and Smith Barney Shearson, Inc., were detailed in "Second Amended Complaint for Violations of the Federal Securities Laws" (pp.42–47) filed by the law firm Gold, Bennet & Cera LLP on March 29, 1996 in San Francisco U.S. District Court.

I have drawn on my coverage of the development of the SMUD wind farm from *California Energy Markets* and the McGraw-Hill newsletters *Electric Utility Week* and *Independent Power Report.*

The details of Kenetech's revenue problems at the time of the IPO offerings were found in the above-referenced story, "Kenetech Pushes New Wind Technology with 6-Million Share Initial Offering," in *Independent Power Report.*

I have drawn on my coverage of the evolution of the BURP for numerous energy trade newsletters, including McGraw-Hill's *Integrated Resource Planning Report;* numerous commentaries of mine on the topic, including the previously referenced "Winds of Change" in the *Sacramento News & Review;* "The Utility Dinosaurs' Power Play," *San Diego Union Tribune,* January 19, 1994, p. B-5; Smeloff and Asmus, *Reinventing Electric Utilities,* pp. 69–70, 73, 74, 75–76, 195–196.

Evidence of the infighting between wind power companies because of the BURP can be found in David Simpson and Jeff Shopoff, "FloWind Corporation's Protest to SeaWest Energy Corporation's Petition for Modification," filed at the California Public Utilities Commission on September 25, 1990, pp. 4–5.

Chapter 21

I interviewed UC-Davis professor Robert Thayer by phone in September 1991 and in the spring of 1994, and gathered further information from his book, *Gray World, Green Heart: Technology, Nature, and the Sustainable Landscape* (New York: John Wiley & Sons, 1994), pp. 273–276; his report "Altamont: Public Perceptions of a Wind Energy Landscape," *Landscape and Urban Planning,* May 1987, pp. 393–397; and his "Wind on the Land," *Landscape Architecture,* March 1988, p. 73.

Much of the material on wind projects in the Pacific Northwest was originally published in my article, "Reaching a Better Understanding," *Windpower Monthly,* January 1996, pp. 35–39; and in the two previously referenced articles of mine in *E Magazine* and *Amicus Journal.*

Information not drawn from previously published works came from:

Testimony of Dennis White before Bonneville Power Administration and Klickitat County regarding Washington Wind Plant #1, April 7, 1995, p. 1.

A September 30, 1995, letter from Jay Letto, president of the Central Cascades Alliance, to Secretary of the Interior Bruce Babbit and Secretary of Energy Hazel O'Leary.

An August 2, 1995, letter from Paul Ketcham, Conservation Director for the Audubon Society of Portland, to Babbit and O'Leary.

The information about the nearby gas-fired power plants was provided by the Portland-based Renewables Northwest Project over the phone in spring 1996.

I interviewed Ron Wiggins by phone in the spring of 1994 while writing my story for *Amicus Journal.* I also used a quote of his that I found in Lauren Picker, "Is the Answer Blowing in the Wind?" *House Beautiful,* September 1992, p. 61.

I compiled the information about Country Guardian from a variety of printed sources provided by James Dehlsen, George Wagner, and Ron Wiggins and obtained from the website http://ourworld.compuserve.com/homepages/windfarms. The Raymond Suitor quote was found in *Country Guardian,* No. 6, October 11, 1993, p. 2.

The information about Minnesota and Texas was derived from interviews I conducted for *Windpower Monthly* and McGraw-Hill's *Utility Environment Report.*

Chapter 22

The letter by Michael Alvarez, vice president of Kenetech Windpower, was addressed to Burl W. Haar, Executive Secretary, Minnesota Public Utilities Commission, and dated July 18, 1995. Information used to develop this chapter can be found on pp. 1–8.

Paul Gipe's e-mail message was sent to John Dunlop, an AWEA representative, on July 2, 1995. Tinkerman forwarded a copy to me on July 26, 1995.

Henry Hermann's recommendation to sell Kenetech stock was published on behalf of WR Lazard, Laidlaw & Mead on August 22, 1994, pp. 2–3.

The *Second Amended Complaint for Violations of the Federal Securities Laws,* filed by George Trevor, Gold, Bennett and Cera on March 29, 1996, was the source of the Kenetech stock price decline data (pp. 2, 25), fees paid to Merrill Lynch and Morgan Stanley (pp. 17, 18), and violations of the federal Securities and Exchange Commission codes (pp. 25, 49). The quote from this report appeared on p. 39, and the account of the mounting financial difficulties in 1993 is on pp. 32–33.

Chapman, Eckland, and Charren supplied information during conversations referenced in prior chapters.

Chapter 23

The opening sequences of this chapter were based on interviews conducted with Dehlsen in the winter of 1997–1998. The specifics about the Zond/SeaWest transmission line were supplied by Robert Kahn, who wrote this episode up for his consulting firm as a case study of a siting success.

I interviewed Ken Karas first via fax (I wrote questions; he gave me written responses) and later by phone, in late April 2000. The details of Karas's career with Zond (later Enron Wind Corp.) were provided by Mary McCann, Enron corporate communications chief.

Ed DeMeo's observations were pulled from a short slide presentation, "What Happened to Kenetech Windpower?" which he created in October 9, 1996, and from a fall 1999 phone interview.

I interviewed Gerald Alderson by phone while still working for *Windpower Monthly* in 1997, to check out a rumor about the re-emergence of Kenetech. I conducted a far more detailed phone interview in September 1999.

The problems with the Kenetech turbines were documented in my story, "Litany of Design Failures," *Windpower Monthly,* September 1997, p. 30.

The Zond test stand information was gathered during a tour of the Zond facility arranged by Dehlsen in preparation for a story written in early spring 1997 for *Windpower Monthly* but never published.

Tinkerman printed out for me his and Stoddard's e-mail exchange during our last in-person interview, in spring 2000.

The performance record of Zond turbines was gathered during my research for the writing of the story "Wind Power Takes Off," *Electric Perspectives,* Vol. 24, No. 26, 1999 (a publication of the Edison Electric Institute), pp. 36–48.

Chapter 24

Virtually all of this chapter has appeared in one form or another in books, reports, and articles I wrote.

The summary of the California energy market in the mid-'90s is based on the book I coauthored with Ed Smeloff, *Reinventing Electric Utilities,* pp. 75–83.

All of the information regarding the birth of California's green power market was previously published in my report, *Wind Energy, Green Marketing and Global Climate Change,* California Regulatory Research Project (a project of the Center for Energy Efficiency & Renewable Technologies), June 1999, pp. 2–10. Much of the same information was also published in "Demanding Cleaner Power," *California Journal,* August 1999, pp. 38–42; "Green Power to the People," *green@work,* January/February 2000, pp. 32–36.

A summary of early support for green power was found in Kari Smith's *Customer Driven Markets for Renewably Generated Electricity,* California Regulatory Research Project (a project of the Center for Energy Efficiency and Renewable Technologies), January 1996, pp. 8–16.

The primary source of the information about Colorado's successes was the report, *Promoting Renewable Energy in a Market Environment: A Community-based Approach for Aggregating Green Demand,* Land and Water Fund of the Rockies/Community Office for Resource Efficiency, May 1997, pp. 1–8, 17–18, 37. I also conducted several in-person, phone, and e-mail interviews with Rudd Mayer.

Osborn, Stoddard, Karas, and Dehlsen interviews round out the rest of the story of how Enron Wind Corp. surpassed Kenetech as the leading American wind power company.

Much of the reporting on the purchase of Zond by Enron Wind Corp. also appeared in my article, "Not Just a Marriage of Marketing Convenience," *Windpower Monthly,* June 1997, pp. 36–39.

Chapter 25

The figures on Navajo revenues and employment rates were culled from the following two documents presented at a 1991 hearing I attended in San Francisco regarding the California Energy Commission's proposed application of environmental values to discourage dirty, out-of-state coal power imports into the California grid:

James Rothwell et al., *Comments of the BHP-Utah International, Inc. on Resource and Planning Assumptions,* November 19, 1991, Docket No. 90-ER-92, p. 1.

Marshall Plummer et al., *Comments of the Navajo Nation on Resource and Planning Assumptions,* November 19, 1991, Docket No. 90-ER-92, p. 1.

The Peterson Zah quotes, and the later material on the Hopi Kachina dance, are drawn from a transcript I developed for the Center for Resource Management's "Why Renewables? Why Now?" conference held in Santa Fe, New Mexico, May 15–16, 1993.

I collected the statistics on coal while performing research for the Center for Energy Efficiency & Renewable Technologies and for the *Power Scorecard,* the latter being a project of the Pace University Energy Project.

The testimonies came from a document presented on October 18, 1999, to David Freeman, president of the Los Angeles Department of Water and Power, by Peabody Coal Company: *Health and Relocation Effects; Testimonies Gathered from Mid-September to mid-October, 1999,* p. 1

The Marsha Monestersky and Gabor Rona quotes, as well as some of the history of the Navajo-Hopi dispute and the environmental impacts of the Mojave power plant, were taken from Victor Megia, "Power Play," *New Times of Los Angeles,* August 8, 1999. The rest of the history came from Malcolm Benally, "To Never Be Seen Again," *Threshold* (Tucson, AZ: Student Environmental Action Coalition, a project of the Tides Center), Winter 1997, pp. 7–8, 11–13.

Most of the remainder of this chapter appeared originally in the following three publications: Peter Asmus, "Native Americans See Money in the Wind," *Windpower Monthly,* July 1996, pp. 26–29; "Landscapes of Power," *Amicus Journal,* Winter 1998, pp. 11–13; and Jim Williamson, "Solar Power for Native Americans," *Clean Power Journal/Special Earth Day Edition,* Center for Energy Efficiency & Renewable Technologies, April 1997, p. 3.

The information about the unique administrative advantages Native American tribes have to offer to the wind power industry was contained in a white paper entitled *Indian Tribes: Their Unique Role in Developing The Nation's Renewable Energy Resources,* Center for Resource Management, January 1997, pp. 16–17.

Chapter 26

Most of this chapter is self-explanatory, but the account of the battle between Enron Wind Corp. and Tejon Ranch Co. and Audubon draws on my article "Birds of a Feather Don't Always Stick Together," published in the on-line magazine *Grist,* January 10, 2000.

Chapter 27

Information about wind power and Kyoto came from two sources: Ros Davidson and Lyn Harrison, "Heading for Kyoto with a United Delegation," *Windpower Monthly,* December 1997, pp. 36–37; and a document Dehlsen put together for AWEA, "Wind Energy and Climate Change: A Proposal for a Strategic Initiative" (1997).

The statistics about small wind turbines came from a slide presentation put together by Mike Bergey.

I interviewed Lawrence Mott during a tour of small wind turbine installations in conjunction with the June 1999 AWEA conference in Burlington, Vermont.

I interviewed Carl Weinberg by phone in November 1999.

I interviewed Oakland Mayor Jerry Brown in April 2000 for a story, "With New Buy, Oakland Will Become World's Largest Green Power City," that appeared in *Currents* (a publication of the Local Government Commission), May/June 2000, pp. 1, 2, 4.

I interviewed Peter Munser by phone in the fall of 1999.

The Egan quote is from his book, *Lasso the Wind: Away to the New West* (New York: Knopf, 1998), p. 239.

Projections of the amounts of increased wind power needed to contain CO_2 emissions are based on calculations I performed with the help of Rich Ferguson, research director for the Center for Energy Efficiency and Renewable Technologies; these appeared in *Wind Energy, Green Marketing and Global Climate Change,* California Regulatory Research Project, pp. 14–15; and "Trends in the Wind: Lessons from Europe and the U.S. in the Development of Wind Power," *Corporate Environmental Strategy,* Vol. 7, No. 1, Spring 2000, pp. 59–60.

Epilogue

This chapter is based largely on interviews conducted during and after the AWEA 1999 conference in Burlington, Vermont, and during subsequent meetings with Tinkerman in the spring of 2000.

Appendix

Most of this material has been previously published in the *Amicus Journal, Grist,* and *Terra Nova.* Much of the information was gathered at meetings I attended of the Kenetech Avian Research Task Force in October 1995, and subsequent interviews with Cade, Hunt, Curry, and Kerlinger.

Some of the original sources include:

Richard Banks, *Human Related Mortality of Birds in the United States,* U.S. Department of the Interior, Fish and Wildlife Service, Special Scientific Report, Wildlife No. 215, Washington, D.C., 1979, p. 14.

Predatory Bird Research Group, *A Population Study of Golden Eagles in the Altamont Pass Wind Resource Area: Population Trends Analysis, 1994–1997,* National Renewable Energy Laboratory, December 1998, pp. 1–2, 8–10.

Predatory Bird Research Group, *A Pilot Golden Eagles Population Study in the Altamont Pass Wind Resource Area, California,* National Renewable Energy Laboratory, May 1995, pp. i–ii, 32–40, 103–104.

Richard Anderson, *Avian Monitoring and Risk Assessment Tehachapi and San Gorgonio WRA,* California Energy Commission [Subcontract No. ZAT-6-15179-02], presented at the 1999 AWEA conference and updated by Anderson for the AWEA conference in 2000.

Gregory Johnson et al., *Avian Monitoring Studies: Buffalo Ridge, Minnesota Wind Resource Area, 1996–1998,* Western EcoSystems Technology, Inc., April 9, 1999.

The information about Green Ridge Services proposal to reduce eagle mortality in the Altamont Pass was gathered at a meeting I attended at Kenetech's Livermore office in the summer of 1998.

Carl Thelander was interviewed at the AWEA 2000 conference.

Index